SpringerBriefs in Geography

SpringerBriefs in Geography presents concise summaries of cutting-edge research and practical applications across the fields of physical, environmental and human geography. It publishes compact refereed monographs under the editorial supervision of an international advisory board with the aim to publish 8 to 12 weeks after acceptance. Volumes are compact, 50 to 125 pages, with a clear focus. The series covers a range of content from professional to academic such as: timely reports of state-of-the art analytical techniques, bridges between new research results, snapshots of hot and/or emerging topics, elaborated thesis, literature reviews, and in-depth case studies.

The scope of the series spans the entire field of geography, with a view to significantly advance research. The character of the series is international and multidisciplinary and will include research areas such as: GIS/cartography, remote sensing, geographical education, geospatial analysis, techniques and modeling, landscape/regional and urban planning, economic geography, housing and the built environment, and quantitative geography. Volumes in this series may analyze past, present and/or future trends, as well as their determinants and consequences. Both solicited and unsolicited manuscripts are considered for publication in this series.

SpringerBriefs in Geography will be of interest to a wide range of individuals with interests in physical, environmental and human geography as well as for researchers from allied disciplines.

Sanchita Daripa • Soumyananda Dinda

Hurdles in Fuel Choice in Rural India

Springer

Sanchita Daripa
Department of Economics
The University of Burdwan
Bardhaman, West Bengal, India

Soumyananda Dinda
Department of Economics, Humanities Building
The University of Burdwan
Bardhaman, West Bengal, India

ISSN 2211-4165 ISSN 2211-4173 (electronic)
SpringerBriefs in Geography
ISBN 978-3-031-71604-1 ISBN 978-3-031-71605-8 (eBook)
https://doi.org/10.1007/978-3-031-71605-8

This Springer imprint is published by the registered company Springer Nature Switzerland AG
The registered company address is: Gewerbestrasse 11, 6330 Cham, Switzerland

Preface

Fuel choice of households and their actual consumption decision are important research agendas for achieving sustainable development goals (SDG) within the target year of 2030. Household's fuel choice and amount of fuel consumption decision might assist to control the risk of uncertainty which is associated with climate change that could be a potential threat to sustainable development in emerging economy like India. It is important to identify the determining factors of fuel choice and consumption decision for cooking purpose at household level. The demand for cooking fuel must be identified at household level under different socio-economic conditions which strongly suggests to redesign energy policies focusing on targeted SDGs. To make an effective energy policy design for achieving SDGs, we have to identify factors influencing household's fuel choice decisions at multiple stages, and here we consider two stage-based decisions—first, selection of fuel type, and second, quantity consumption decision. Households take decision at each stage.

Previous studies mainly focused on fuel price and household income determining domestic fuel consumption, and least attention has been given on the stage-based fuel choice decision at household levels in rural area. Again, methodologies used in earlier studies are limited to Probit, Logit and multinomial logit models only when fuel choice is concerned, while Tobit model is taken when both fuel choice and quantity demanded are considered. These methodologies were based on a restrictive assumption that if desired consumption is negative, then actual consumption is zero, but if desired consumption is positive, then actual consumption is equal to desired consumption. This assumption might not always hold. The actual consumption might be zero in spite of a positive desire to consume. This group of observations whose actual consumption is zero in spite of a positive desire to consume are neglected in the methodologies used earlier as mentioned in the literature. Neglecting these observations might not reveal the true fuel consumption scenario at household level in rural India. To overcome this limitation, the double hurdle model (Cragg, 1971) is used in this study to estimate the hurdles in fuel demand and its determinants at the household level in rural India.

This book has 4 parts containing 12 chapters. Part I provides conceptual idea, related literature and data description and methodology. Part II depicts the distributional pattern, Part III provides estimated results of socio-economic determinants and interprets the findings, and finally, Part IV analyses and discusses estimated results of specific categories including castes in rural India.

We are grateful to Dr Sandeep Sarkar for his suggestions for application of Cragg's Double hurdle model in case of fuel choice and consumption. We are grateful to Professor Arpita Ghosh for her comments and suggestions on the earlier draft. We are also grateful to the conference chairs and participants for their comments at Jadavpur University, Kolkata and Flame University, Pune.

Bardhaman, West Bengal, India

Sanchita Daripa
Soumyananda Dinda

Contents

Part IV Analysis

About the Authors

Soumyananda Dinda is working as a Professor of Economics at the Department of Economics, University of Burdwan, India. Previously, he taught at several institutes and universities like University of Bath, UK; Madras School of Economics in Chennai; Chandragupt Institute of Management at Patna; S.K.B. University at Purulia; and S.R.F. College, Beldanga, Murshidabad. He also taught as guest faculty at Presidency College and University of Calcutta. He also worked as Economic Affair Officer in the UN ESCAP, Bangkok. He has visited many foreign institutes and universities like International Centre for Theoretical Physics (ICTP), Trieste; Peace Institute, Norway; Nottingham University; Reading University; Oxford University; Birmingham University; Norway State University; and Prertoria University. He started his academic journey as a Research Fellow at Indian Statistical Institute, Kolkata. He is working on green economics, climate change and technology and different social aspects. He has published several papers in international journals and also published four books.

Sanchita Daripa was an M.Phil. scholar in the Department of Economics at the University of Burdwan, India. She is teaching in school and working on fuel choice at household level. She has worked at NSSO for short time and learnt about NSSO and its data collection. She is an expert in Stata software. She has participated in several conferences, seminars and workshops. She has published one journal article and two book chapters.

Abbreviations

DHM	Double hurdle model
J&K	Jammu and Kashmir
LO	Land own
LPG	Liquified petroleum gas
MPCE	Monthly per capita consumption expenditure
NSSO	National Sample Survey Office
OBC	Other backward classes
SC	Scheduled caste
SDG	Sustainable Development Goals
ST	Scheduled tribes
TN	Tamil Nadu
UP	Uttar Pradesh

Part I
Introduction of Conceptual Concepts and Ideas

Chapter 1
Introduction

1.1 Introduction

Energy consumption is essential for economic activities, and energy requirements are increasing at a high rate along with economic development. In a developing economy, like India, a large part of total energy consumption is contributed by the energy consumption of domestic sector. Household energy demand comes basically from their fuel consumptions. Fuel is an essential item of daily consumption basket of households. Household consumes fuel to serve three basic domestic purposes—cooking, lighting and space heating.[1] Energy consumption for the purpose of cooking is one of the basic requirements for most of the households in India. The principal sources of fuels for the purpose of cooking often turn out to be solid fuels like firewood and chips, charcoal, coke, coal, dung cakes, etc., which cause many health-related diseases.[2] Out of the huge set of solid biomass fuels in this study, we focus on the consumption of firewood and chips. This study mainly focuses on rural India where more than 65% households till now use firewood and chips as the principal source of cooking fuel (DeFries & Pandey, 2010).

The domestic purposes are served by a variety of sources of fuels, namely coal, petroleum, LPG, electricity, biogas, fuel oil, charcoal, dung cakes, firewood and chips, etc. Some of these fuels are traditional fuels like coal, charcoal, dung cakes,

---

[1] Mainly, firewood is used for room heating during winter season in North India.

[2] Burning of these fuels is the main source of indoor air pollution, which is linked to a range of health issues, e.g. increase in the risk of chronic obstructive pulmonary disease and acute respiratory infections in childhood. Following Smith and Mehta (2003), indoor air pollution has caused approximately two million annual premature deaths in developing countries. Furthermore, following the estimate of Smith, 400–500 thousand premature deaths in India can be attributed annually to the use of biomass fuels. See Heltberg (2005) for further details on diseases related to air pollution.

S. Daripa, S. Dinda, *Hurdles in Fuel Choice in Rural India*, SpringerBriefs in Geography,
https://doi.org/10.1007/978-3-031-71605-8_1

crop residues, firewood and chips, etc. The efficiency of traditional fuels is lower than modern fuels like LPG, electricity, etc. Apart from the efficiency standards, combustion of the traditional fuels creates pollution, and they are also termed *dirty fuels*. Their combustion also leads to several health hazards. Modern fuels like LPG, biogas, electricity, etc. are termed *clean fuels* which are more efficient than traditional fuels. Till today, dirty fuels are the major source of cooking fuels for rural households in developing economies like India. So, household fuel choice decision, in India, is an important aspect of the study.

1.2 Sustainable Development Goals (SDGs)

Dependence on firewood items is also responsible for deforestation and forest degradation. Forest is often considered a renewable source of energy (Couture et al., 2012), while firewood is considered dirty fuel since it adds to CO_2 emissions upon combustion. In developing countries, a large part of the supply of firewood comes from deforestation, which also acts as a threat to the environment. Developing economies are ushered into a world of uncertainty with unsustainable development. In this context, international organisations like the United Nations (UN) play a crucial role. The UN has set up universal goals in the framework of Sustainable Development Goals (SDGs) which is the extension of the Million Development Goals (MDGs) that were set up in the early twenty-first century. The MDGs mainly focus on the development of human capital while SDGs set a wide coverage on human capital formation and its protection for sustaining humankind. Now, the global human society feels an urgent need for interconnections among economic, social and environmental dimensions of development. The basic core objective of SDGs is to make sustainable consumption (Akenji & Bengtsson, 2014). So, an urgent social need is to design policy and promote sustainable consumption, which focuses on non-polluting consumption; the process of production and consumption should minimise pollution. The indicators of sustainable consumption are directly or indirectly connected with energy consumption, which should be attributed to clean energy consumption in civil society. Recently, sustainable consumption[3] and production patterns have gained importance in policy dialogues at the international level (Gasper et al., 2019; Ma et al., 2019; Xu et al., 2018; Scherer et al., 2018). Goal 12 (SDG12) ensures a sustainable consumption and production process, which is interconnected with Goal 7 (SDG7) focusing on clean energy. Entire socio-economic development evolves over production and consumption patterns which should ensure sustainability of this consumption and production process over time unless and until improving energy efficiency and/or reducing dirty fuels. Sustainable consumption encourages socio-economic development within possible market

[3] Sustainable consumption suggests to fulfil the basic needs and improve quality of life without harming the environment, economy and society overall.

ecosystems. It reveals the consumer's choices of goods and services focusing on basic needs, which actually reflect the demand side economy for quality consumption to sustain the long run.

In this context, it is required to understand the determinants of domestic fuel choice decisions at the household level with their social and economic constraints. Fuel choice of households and their actual consumption decision are important research issues for achieving Sustainable Development Goals (SDG) within 2030. Households' fuel choice and amount of fuel consumption decision may certainly help to control the risk of uncertainty associated with climate change which could be a potential threat to sustainable development in emerging economies like India. So, it is important to identify the determining factors of cooking fuel choice decisions at the household level under different socio-economic conditions that definitely help to reformulate and redesign energy policy focusing on targeted SDGs. To make an effective energy policy design for achieving SDGs, we have to identify factors influencing households' fuel choice decisions at multiple stages.

1.3 Research Issue

Earlier studies[4] have mainly focused on income- and price-determining domestic fuel consumption, and little attention has been given to the stage-based fuel choice decision at household levels in developing economies like India. In this context, considering the decision-making process in detail at different stages of choice and consumption, a study is required to re-examine the determinants of domestic cooking fuel choice in India and more specifically in rural area that covers a large population. Suppose some households have a positive desire to consume as well as make a positive amount of consumption, while some other households have a positive desire to consume, but their actual consumption is zero due to certain factors that inhibit carrying out of the change for actual consumption to be positive. So, there is a difference between desire and actual demand which is also mentioned in economic theory. There might be some households having a positive desire to consume firewood, but the actual demand of firewood is zero. However, these households are ignored in earlier literature having Tobit[5] model which is used only for positive demand and actual demand, given that actual demand is positive. Positive demand and actual demand generate in two different processes, whereas Tobit model's outcomes are determined by the same process. This is the main limitation of Tobit model for corner-solution. Hence, this study incorporates Cragg's two-stage (or double hurdle) model alternative to Tobit model for corner-solution. It should be noted that one underlying process determines Tobit model's outcomes while

[4] See Couture et al. (2012), Hanna and Oliva (2015), Heltberg (2005), Jingchao and Kotani (2012), etc.

[5] Tobin (1958) developed a model on limited dependent variables. It is popularly known as Tobit model.

outcome of two different processes is Cragg's (1971) double hurdle model. Cragg (1971) model is more flexible allowing two separate processes (i) probit model in the first tier and (ii) truncated normal model in the second tier. Truly, Cragg (1971) model integrates the probit model to determine the probability of positive demand and truncated normal model for the given positive demands. So, the double hurdle model of Cragg (1971) is more general version of Tobit model.

In this context, this study investigates hurdles in the selection of firewood and chips as cooking fuel and identifies stage-based determinants of its choice and consumption decision at household level in rural India. Applying the double hurdle model of Cragg (1971), this study re-examines the determinants of domestic fuel consumption focusing on hurdles in the decision-making process for the selection of firewood and chips as a cooking fuel and its actual consumption decision in rural India. Household-level fuel consumption primarily depends on the socio-economic background of the households, not on the market. The market for traditional fuels is underdeveloped or absent in rural India. So, price mechanism may not be a crucial determining factor of fuel choice in rural area of India. Various studies (Gupta & Kohlin, 2006; Gundimeda & Kohlin, 2008; Jingchao & Kotani, 2012) have pointed out several determining factors of fuel consumption at the household level like income, age, sex, education, caste, religion, etc. Apart from these studies, some other studies (Barnes et al., 2010) have pointed out accessibility and ease of use of fuels as important determining factors for fuel consumption choice. In spite of this, there are several studies on determinants of domestic fuel consumption, which is possibly because the mere presence of resources is not the only reason for its utilisation or consumption. Literature have pointed out several determining factors of fuel consumption at the household level like income, religion, caste, sex of the head of the household, age of the head of the household, educational level, etc.

1.4 Research Gaps

Most of the studies on fuel choice decisions are conducted across various countries of the world. However, almost all the studies were focussed towards the demand for fuel at the household level, with little focus on the supply side. In a developing country like India, the supply of fuel, i.e. the fuel source, might play an important role in making several policy decisions regarding fuel choice and amount consumption at the household level. For example, we consider firewood as a cooking fuel, which may be purchased from the market or home-grown for fuel use or it is collected freely from neighbouring areas of the household.

Apart from this, as mentioned above, the methodologies used in earlier studies are mainly limited to multinomial logit model (Couture et al., 2012; Takama et al., 2012), multinomial probit model (Jumbe & Angels, 2011; Gupta & Kohlin, 2006) when only fuel choice is concerned and Tobit model (Jingchao & Kotani, 2012) when both fuel choice and quantity demanded are taken. However, these methodologies were based on a restrictive assumption that if desired consumption is negative, then actual consumption is zero, but if desired consumption is positive, then actual consumption is equal to desired consumption. However, this assumption might not hold always. It may be possible that desired consumption is positive, but actual consumption is not so because other factors inhibit carrying out the change for actual consumption to be positive, i.e. there exists a distinction between desire and effective demand. So, in spite of a positive desire to consume the actual consumption might be zero. This group of observations whose actual consumption is zero in spite of a positive desire to consume are neglected in the methodologies used earlier as mentioned in literature. Neglecting these observations might not reveal the true scenario. To overcome this limitation, the double hurdle model[6] proposed by John G Cragg (1971) is used in the current study to estimate the hurdles in fuel demand and its determinants at the household level in rural India.

1.5 Objective of the Study

The main objective of the study is to analyse the determinants of firewood demand for cooking purposes in rural India. In light of the main objective, the specific objectives of the study are as follows:

- To understand the distributional pattern of fuel consumption at the household level in rural India.
- To examine the determinants of consumption of firewood and chips, the most preferred fuel, in rural India.
- To analyse the firewood and chips demand for cooking purposes at the bottom and top quartile income groups in rural India study.
- To analyse the choice of cooking fuel across castes in rural India.
- To examine the determinants of firewood and chips consumption of female-headed households in rural India.

1.6 Plan of the Study

This study focuses on hurdles of cooking fuel choice and its consumption, and the study is organised as follows: Chap. 2 reviews literature. Chapter 3 describes data and methodology for analysis purpose. Chapter 4 provides distributional patterns of fuel consumption in rural India. Chapter 5 discusses the distributional pattern of firewood and chips in rural India. Chapter 6 determines and identifies the factors responsible for the choice of fuel consumption. Chapter 7 analyses stage-based determinants of non-price factors of firewood and chips choice and its consumption. The double hurdle model is applied here for finding factors at two different stages.

[6] It was originally used to estimate the demand of durable goods.

Chapter 8 provides determining factors of firewood and chips selection and consumption in the largest populated state, Uttar Pradesh (UP). Chapter 9 shows the determinants of firewood and chips consumption for top and bottom quartile groups. Chapter 10 discusses the choice of firewood and chips across castes in rural India. Chapter 11 shows the hurdles of female-headed households for cooking fuel choice and consumption in rural India. Finally, Chap. 12 concludes with remarks.

Chapter 2
Review of Literature

2.1 Introduction

Cooking is one of the major activities of a household in a developing country. The choice of cooking fuel is an important decision at the household level. Generally, it is expected that the cooking fuel choice decision of households depends on the socio-economic background of the households. But the case might not be as simple as it is observed. It might be possible that despite vast endowments of energy or fuel sources, countries might lag behind in energy production as well as energy consumption like sub-Saharan Africa (Kebede et al., 2010). There are several studies on cooking fuel choice decision of households. The demands of cooking fuels are determined by several economic factors as well as social factors. It is thus, required to understand the determinants of fuel consumption at the household level and estimate the cooking fuel demand. Studies on cooking fuel demand can be broadly classified into two major groups, namely theoretical literature and empirical literature.

2.2 Theoretical Literature

Existing theoretical literature are mainly of three types, namely energy ladder hypothesis, fuel stacking models and improved cookstove models.

S. Daripa, S. Dinda, *Hurdles in Fuel Choice in Rural India*, SpringerBriefs in Geography,
https://doi.org/10.1007/978-3-031-71605-8_2

2.2.1 Energy Ladder Theory

The study of literature reveals that with lower levels of economic development, households tend to consume more traditional fuels like dung cake, coal, firewood, etc., which are possibly cheaper. These fuels create pollution upon combustion, and so they are called dirty fuels. As economic development takes place, they are replaced by modern or cleaner fuels like LPG, electricity, etc. Thus, with an increase in the level of economic development, households have a tendency to discard traditional or dirty fuels, and modern or clean fuels are accepted. This fuel-switching pattern from dirty to clean fuels is called the energy ladder hypothesis (Leach, 1992). Households rank different available alternative fuels in order of their preference based on physical characteristics of fuel like cleanliness, ease of use, cooking speed, etc. and switch from the least preferred to the most preferred fuel with an increase in income or level of economic development. This is very much similar to the idea of climbing higher steps of a ladder and perhaps that is why this theory is called the energy ladder hypothesis. Several studies have been made focusing on the energy ladder hypothesis (Gupta & Kohlin, 2006; Gundimeda & Kohlin, 2008; Barnes et al., 2010; Davis, 1998; Hosier & Dowd, 1987; Pachauri & Jiang, 2008).

Though most of the studies have supported the energy ladder hypothesis, still there are some who have refuted the idea of energy ladder hypothesis. Hanna and Oliva (2015) assessed the impact of an increase in the economic well-being of households through an asset transfer program on fuel choice decision focusing on fuel expenditure and fuel composition in the Murshidabad district of West Bengal. The data used in this study is from a randomised experiment of an asset transfer programme in the Murshidabad district of West Bengal administered by Bandhan, a local NGO. Eight hundred households received livelihood intervention which consisted of a grant in the form of an asset (cow or goat) and a stipend depending on their asset. It was observed from the study that improvement in well-being after the programme also led to an increase in fuel consumption of the households, but there was no shift towards cleaner fuels for cooking. In fact, households resorted to inferior fuels for cooking, i.e. consumption of cow dung increased which is possibly because of its readily available nature through the asset transfer programme.

2.2.2 Fuel Stacking Theory

The energy ladder hypothesis assumes a complete substitution of dirty fuels by clean fuels. But some studies suggested that complete substitution of dirty fuels by clean fuels is not possible as suggested in the energy ladder hypothesis. In fact, sometimes households continue to consume more than one fuel. They tend to accept modern or cleaner fuels but cannot completely discard the traditional fuels, i.e. incomplete substitution of dirty fuels takes place. This is called the fuel stacking model (Masera et al., 2000) where households continue to consume more than one fuel simultaneously. In fact, many have referred that multiple fuel use is more common (Couture et al., 2012; Arnold et al., 2006; Brouwer & Falcao, 2004; Campbell et al., 2003). Studies have shown that firewood which is considered a dirty fuel displays the characteristics of a normal good when used as backup fuel, but it shows the characteristics of an inferior good when used as main fuel, i.e. consumption of firewood as the main fuel decrease with increase in income, thereby supporting energy ladder hypothesis, but it is not completely substituted by cleaner fuels and continue to be used as backup fuel with increasing income of households which supports the fuel stacking models (Couture et al., 2012).

2.2.3 Improved Cooked Stove Models

Apart from the energy ladder and fuel stacking models, another most widely used strategy to reduce biomass fuel consumption in developing countries is the distribution of improved cookstoves. These studies were mainly focused on studying the impact of various product-specific characteristics like usage cost, safety risks, etc. (Takama et al., 2012) or contextual factors like internal or external household environment (van der Kroon et al., 2014) and how they affect the adoption of improved cookstoves to reduce biomass fuel consumption. Though improved cookstove may perform better in controlled conditions, the results in real situation may differ. For example, adopters may switch to more firewood-intensive meals using the new technology or differences in education, preference wealth, etc. may fail to use stoves as it was used under controlled conditions. Further field study compares the firewood consumption of users and non-users and these two groups may differ in terms of ability health and preferences amongst them which is called the self-selection issue. Adrianzen (2013) has addressed this self-selection issue in his paper and investigated the "on the field" impact on firewood consumption of an improved stove design. Raman (2014) also stated that the major challenge is to narrow down the gap between the test results obtained under laboratory conditions and in the real world. He stated that though most households rely on traditional stoves for cooking, these stoves have low efficiency and cause pollution and health problems. Several tests are used to evaluate the performance of these traditional stoves. His paper aimed to reduce the gap in test methodology in such a manner that results obtained in the laboratory are comparable with results of actual cooking. Common errors during water boiling tests are identified, and such errors are also proposed to be minimised. Banerjee (2016) aims to check if the dissemination of induction stoves in Himachal Pradesh influences the existing household fuel consumption patterns and how much feasible is electricity as a domestic clean cooking fuel. The results revealed that electricity replaced LPG as a secondary fuel but failed to replace traditional mud stoves as the primary cooking technology. It was also seen that a shift from firewood to electricity was observed for only 5% of household and those falling in the low monthly per capita consumption expenditure (MPCE) category have lesser access to electricity. On the whole, the results indicate electricity has limited potential to reduce firewood usage.

2.3 Empirical Literature

Apart from the theoretical studies, there are certain empirical works also in the study of household cooking fuel demand. They can be summarised as follows: The empirical literature on cooking fuel demand of households can be classified under the following groups.

2.3.1 Economic Factors

The most important economic factor affecting fuel choice is the level of income. Almost all literature support the role of income in affecting the fuel choice decision of households (Gupta & Kohlin, 2006; Gundimeda & Kohlin, 2008; Couture et al., 2012; Jingchao & Kotani, 2012). It is often considered as a proxy variable for the level of economic development. The energy ladder theory is primarily based on the notion of income indicating an increased income leads to a shift to cleaner fuel consumption (Leach, 1992).

2.3.2 Social Factors

Following the energy ladder hypothesis, some other studies have highlighted the importance of non-income factors or social factors like the age of the head of household, family size, caste, religion, education, etc. in determining the fuel choice of the household.

2.3.2.1 Family Size

Family size is an important determining factor of household fuel consumption. Studies have shown that household size has negatively affected the probability of adopting cleaner fuels, i.e. larger families have a tendency to consume more biomass fuels like firewood (Gundimeda & Kohlin, 2008; Jumbe & Angelsen, 2011; Jingchao & Kotani, 2012; van der Kroon et al., 2014).

2.3.2.2 Education

Literature suggests that educated households have a higher tendency to consume clean fuels than uneducated households who prefer biomass fuels or traditional fuels, i.e. increased level of education of the households encourages a transition towards cleaner fuels (Jingchao & Kotani, 2012; Gebreegziabher et al., 2012).

2.3.2.3 Caste System

It is also observed that though caste system is officially abolished, households of backward classes tend to consume more dirty fuels than other castes (Gupta & Kohlin, 2006; Heltberg, 2005; Hosier & Dowd, 1987; Ouedraogo, 2006).

2.3.2.4 Gender

Several studies have also pointed out the role of women in cooking fuel choice decision of households. Some have mentioned that educated women–headed households or working women–headed households have a tendency to accept more clean fuels (Gupta & Kohlin, 2006), while other studies have showed that women value clean fuels less than men, possibly because the household spending decision lies under a man's domain (van der Kroon et al., 2014).

2.3.2.5 Social Trust

Goswami (2017) studies how the transition to clean cookstoves is driven by other latent factors like caste, trust, social capital, information flows, etc. focusing on eight villages in the Purnia district of Bihar. The findings reveal that caste, socio-political position, asset, remoteness, culture and technology access affect the transition to clean fuels for cooking. These findings enable policymakers in decision-making and help in developing better policy for energy transition in rural India.

2.3.3 Geographical Factors

Apart from these, some studies have highlighted the importance of geographical factors like nearness to forests, etc. in explaining the fuel choice decision of households. (Jingchao & Kotani, 2012; Jumbe & Angelsen, 2011).

2.3.4 Product-Specific Factors

Both the energy ladder and fuel stacking model fail to fully explain the role of consumer choices in fuel and stove selection. The socio-economic factors cannot fully explain the fuel and stove choice decision because these factors change over time. So, a study of product-specific factors across income or wealth groups helps in a better understanding of household fuel and improved clean cookstove choice decisions (Takama et al., 2012; van der Kroon et al., 2014). It is seen from these studies that the importance of product-specific characteristics like stove price and usage cost decreases to determine stove choice as income increases. But other product-specific characteristics like safety risk are highly significant across all wealth groups. The marginal willingness to pay for safer fuel is also high across all wealth groups (Takama et al., 2012; van der Kroon et al., 2014). Internal household characteristics like traditional practices, etc. and household external characteristics like abundance and lack of resources, etc. also affect the fuel choice decision of households. Respondents have a positive attitude towards improved cookstove, but firewood still remains an alternative, and its continued usage is expected representing the fuel stacking behaviour of households. Biomass abundance in certain regions limits the tendency to accept improved cookstoves, but increased market access on the other hand facilitates such improved cookstoves (van der Kroon et al., 2014). Nayak (2015) reveals that households use firewood or kerosene as major cooking fuel and spend as much as LPG, but are unwilling to take up LPG as major cooking fuel due to its high initial investments. So, a community-based cooperative pilot project was taken where all households were given LPG connection by pooling financial resources from the community. It was seen from the pilot project that communities can themselves come up with innovative ideas for achieving transition to clean fuels and thereby achieve sustainable development goals.

2.3.5 *Empirical Models*

Most of these studies are conducted across various countries of the world. The models used in these studies are mainly limited to the multinomial logit model when only fuel choice is concerned (Couture et al., 2012; Takama et al., 2012), multinomial Probit model (Jumbe & Angelsen, 2011; Gupta & Kohlin, 2006) and Tobit model (Jingchao & Kotani, 2012) when both fuel choice and quantity demanded are taken.

2.4 Conclusion

The literature reviewed in the above section were focused towards the demand for fuel at the household level, with little focus on the supply side. In a developing country like India, the supply of fuel, i.e. the fuel source, might play an important role in making several policy decisions regarding fuel choice at the household level. For example, if we consider the fuel say firewood and chips, it may be home grown for fuel use or collected for free from neighbouring areas of the household, or purchased from the market. In these three cases, the policies for encouraging households to switch to a cleaner fuel will be different, which could be taken into consideration only when we consider the source of supply of firewood and chips. The models used in these studies were based on a restrictive assumption that if desired consumption is negative, then actual consumption is zero, but if desired consumption is positive, then actual consumption is equal to desired consumption. However, this assumption might not hold always. It may be possible that desired consumption is positive, but actual consumption is not so because other factors inhibit carrying out the change for actual consumption to be positive. Literature has considered zero consumption considering actual zero consumption added with negative desire consumption converted to zero. To overcome such limitation, the double hurdle Cragg model (1971), which was originally used to estimate the demand for durable goods, is used in the current study to estimate the cooking fuel demand and its determinants at the household level.

Chapter 3
Data and Methodology

3.1 Data Description

The data used for this study is secondary data which is collected and published by the National Sample Survey Office[1] (NSSO), the Government of India. The NSSO conducts nationwide sample surveys relating to various socio-economic topics to collect data for planning and formulation of policies. The socio-economic surveys are conducted in the form of rounds. Each round is normally of one year duration and occasionally for a period of six months. The first round of NSSO was conducted during 1950–51. As per the present practice, the survey period coincides with the agricultural year and starts from July each year. The NSSO 68th round Socio-Economic Survey data which is collected during the period of July 2011 to June 2012 is used in the current study. This data is published in September 2015.

The 68th round (July 2011–June 2012) of NSSO is earmarked for surveys on 'Household Consumer Expenditure' and 'Employment and Unemployment'. The current survey is a quinquennial survey as far as subjects of enquiry, design, questionnaires and sample sizes are concerned. This is the last published quinquennial survey data of the NSSO.

From the data, it is observed that in India, individuals are faced with ten major alternatives for cooking purposes, namely coke and/or coal, firewood and chips, LPG, gobar gas, dung cakes, charcoal, kerosene, electricity, others and no cooking arrangements. Using Stata software, we have performed a tabulation which reflects that out of these ten types of cooking fuel, rural households tend to prefer firewood and chips as the most preferred fuel, and urban households are inclined more towards LPG, i.e. they consider LPG as the most preferred fuel. Table 3.1 shows that 62.68% of households in India consume firewood and chips as the most preferred fuel in rural areas, while 68.9% of households preferred LPG as prime cooking fuel

[1] Earlier it is known as National Sample Survey Organisation.

S. Daripa, S. Dinda, *Hurdles in Fuel Choice in Rural India*, SpringerBriefs in Geography,
https://doi.org/10.1007/978-3-031-71605-8_3

Table 3.1 Distribution of households as per sources of cooking fuels in India

	Rural			Urban		
Cooking fuels	Freq.	Percentage	Cum.	Freq.	Percentage	Cum.
Coke, coal	657	1.1	1.1	914	2.18	2.18
Firewood and chips	*37,410*	*62.68*	*63.78*	*7483*	*17.83*	*20.01*
LPG	14,562	24.4	88.18	28,912	68.9	88.91
Gobar gas	141	0.24	88.42	6	0.01	88.93
Dung cake	4203	7.04	95.46	520	1.24	90.17
Charcoal	25	0.04	95.5	115	0.27	90.44
Kerosene	619	1.04	96.54	1907	4.54	94.99
Electricity	77	0.13	96.67	245	0.58	95.57
Others	1669	2.8	99.46	313	0.75	96.32
No cooking arrangement	320	0.54	100	1546	3.68	100
Total	59,683	100		41,961	100	

Source: Daripa and Dinda (2022)

Table 3.2 Frequency distribution of households using firewood and chips by sources in India

Sources of firewood	Frequency	Percentage
Only purchase	14,562	30.05
Only home-grown stock	13,347	27.54
Both purchase and home-grown stock	1580	3.26
Only free collection	16,631	34.32
Only exchange of goods and services	93	0.19
Only gifts/charities	89	0.18
Others	2160	4.46
Total	48,462	100

Source: Daripa and Dinda (2022)

in urban India. From Table 3.1, it is noted that the second and third most preferred fuels are LPG (24.4%) and dung cake (7.04), respectively, in rural India, while these are firewood and chips (17.83%) and kerosene (4.54) in urban India.

Firewood and chips consumed in rural India come from various sources. Table 3.2 describes the distribution of sources of firewood and chips in rural India. Table 3.2 shows that the main source of firewood and chips is free collection (34.32%) followed by purchase (30.05%) and home grown (27.54%) in rural India.

Our main focus is on the choice of firewood and chips in rural India and identifying determining factors which are considered one by one. The most important determinant of fuel choice of households is income, which is not available. So, following literature, we have considered monthly per capita consumption expenditure (MPCE) as a proxy variable for income. We have discussed the income distribution of households in rural India in the next chapter.

Apart from income, the major asset of rural households is considered as land holding. We have also considered the land-holding pattern of households focusing on land owned by households. We have data on land holdings for only 25,794 households out of total 59,683 rural households. This implies that 33,889 households are landless. The average land holding of 25,794 households is around 0.0102 hectares.

Possibly, social factors govern the fuel choice of individuals such as social groups, religion, gender and education level. Table 3.3 provides descriptive statistics of the social variables. The sex of the head of household plays a major role in determining the fuel choice of households. In the data, the majority are male-headed households who take consumption decision at the household level. The data on education level reflects the educational attainment of the respondents and their members. The data on religion represents eight major religions, namely Hinduism, Islam, Christianity, Sikhism, Jainism, Buddhism, Zoroastrianism and others. The data on social groups represents four major social groups in India, namely ST, SC, OBC and others. This others category represents the general category.

Table 3.3 shows that 89.31% households are male headed and 10.69% are female headed. It is observed from Table 3.3 that most of the household heads in our

Table 3.3 Descriptive statistics of socio-economic variables

Category	Variables	Frequency	Percentage
Sex of the head of households	Male	53,304	89.31
	Female	6379	10.69
Educational level of the head of households	Illiterate	17,354	29.08
	Through EGS/NFEC/AEC	192	0.32
	Through TLC	43	0.07
	Others	171	0.29
	Below primary	7022	11.77
	Primary	8181	13.71
	Middle	9804	16.43
	Secondary	7737	12.97
	Higher secondary	4230	7.09
	Graduate	3437	5.76
	Post-graduate and above	984	1.65
Religion	Hinduism	45,545	76.32
	Islam	7092	11.88
	Christianity	4273	7.16
	Sikhism	1344	2.25
	Jainism	66	0.11
	Buddhism	696	1.17
	Zoroastrianism	1	0.00
	Others	661	1.11
Social groups	Scheduled tribes (ST)	9930	16.64
	Scheduled castes (SC)	10,193	17.08
	Other backward classes (OBC)	23,546	39.46
	Others (general caste)	16,005	26.82

Source: Daripa and Dinda (2020, 2022)

studied population are illiterate (i.e. 29.08%), 11.77% are educated below primary level, 13.71% primary level, 16.43% middle level, 12.97% secondary level and very few are highly educated (7.41% are graduate and above). It is also noted that majority of the population belong to Hindu category (76.39%) followed by Islam (11.88%) and Christianity (7.16%), and all other religions consist of a very small proportion of total population. So, for simplicity, we club these religions into a single group named others. Table 3.3 displays that a majority of the population in our study belong to other backward class (OBC category 39.46%) followed by generals or others category (26.82%), scheduled castes (SC category 17.08%) and scheduled tribe (ST category 16.64%).

3.2 Methodology

We shall now consider an empirical model to explore the role of determinants in the consumption of firewood and chips. The dependent variable nonetheless is limited, in the sense that a household may decide not to use firewood. For such households, entries of dependent variables, essentially 0, otherwise 1. Literature mainly considers logit and probit models for binary or limited dependent variables. Now, we shall address this issue following a double hurdle model of limited dependent variables that has been introduced by Cragg (1971). It may also be considered as a two-step method. The first step is to estimate the probability of making a positive (>0) consumption. The second step is to estimate the intensity of consumption. Both steps identify the role of those factors in estimating cooking fuel choice and consumption. Truly, households have certain constraints and face difficulties in making decisions on the selection of fuel and its actual use or consumption. The decision of fuel choice of household has basically two stages—the first stage is the decision of participation, and the second stage is the decision of quantity or amount of selected fuel consumption.

If the decision to consume fuel only is to be considered, then the dependent variable becomes binary, i.e. 1 and 0 for consuming and not consuming fuel, respectively. In such cases, the methodologies that are used are limited to logit and probit models. The logit and probit models are generally used for choice decisions or binary response models. These models are also used in the case of binary dependent variables, but the major difference between these two models is that the cumulative distribution function (CDF), which provides the shape of the distribution. Logit model follows the logistic distribution while Probit model follows the normal distribution. The logistic distribution has a similar shape to the normal distribution, but with heavier tails.

3.2.1 Tobit Model

However, when the amount of a particular fuel, say firewood consumption, is considered and not merely the choice is considered, then these selection models become rather inappropriate. For example, if we want to examine the amount of firewood consumed by say 100 households in any area and estimate the determinants of firewood consumption, then the earlier selection models become inappropriate. This might be because there might be, say, 60 households who do not consume firewood and other 40 households who consume firewood. In such situations, we have information about only the independent variables for 60 households and information on both the dependent and independent variables for 40 households. In such situation, a tobit model is used. The tobit model considers those observations that report a positive amount of consumption only, ignoring those that report zero consumption.

The tobit model (1958) was developed by Tobin to analyse censored dependent variables. Tobin's model is written as follows:

$$\begin{aligned} y_i &= y_i^* \quad \text{if } y_i^* > 0 \\ &= 0 \quad \text{otherwise} \end{aligned} \tag{3.1}$$

where y_i^* is the latent dependent variable.

The latent dependent variable y_i^* is described by the regression equation:

$$y_i^* = x_i\beta + u_i, \text{where } u_i \sim N\left(0, \sigma_u^2\right) \tag{3.2}$$

u is the independent, identical, homoscedastic and normally distributed disturbance term in the econometric model in Eq. (3.2).

3.2.2 Two-Stage Decision

The tobit model considers those households who have a positive desire to consume as well as makes a positive amount of consumption and ignores those households whose actual consumption is zero. But it might be possible that there exist some households who have a positive desire to consume, but their actual consumption is zero. This might be because there may be certain factors that inhibit carrying out of the change for actual consumption to be positive. In this connection, the difference between *desire and actual demand* is to be noted as mentioned in economic theory. There might be some households having a positive desire to consume firewood, but actual demand of firewood is zero. However, these households are ignored in the tobit model. An alternative model is proposed by Cragg (1971), i.e. the double hurdle model is applicable.

3.2.3 *Double Hurdle Model*

Applying the Cragg (1971) model, this study investigates the determinants of firewood consumption in rural India. Cragg's double hurdle model was initially used to estimate the demand of durable goods. Actually, the double hurdle model of Cragg (1971) is an extension of Tobin's model for limited dependent variable.

Cragg (1971) put forward a two-part extension to Tobin's model in his research paper. He addressed the presence of two latent variables through two regression equations, i.e. participation decision and actual consumption or expenditure decision. The occurrence of these two latent variables actually explains the double hurdle mentioned in the Cragg model.

The latent variables are represented by two regression equations as follows:

$$y_{i1}^{*} = w_i\alpha + \epsilon_1 \tag{3.3}$$

$$y_{i2}^{*} = x_i\beta + \epsilon_2 \tag{3.4}$$

The Cragg model is represented as:

$$\begin{aligned} y_i &= x_i\beta + \varepsilon \quad \text{if } y_{i1}^{*} > 0 \text{ and } y_{i2}^{*} > 0 \\ &= 0 \quad \text{otherwise} \end{aligned} \tag{3.5}$$

where y_{i1}^{*} is a latent endogenous variable representing an individual or household participation decision; y_{i2}^{*} is a latent endogenous variable representing an individual or household consumption or expenditure decision; y_i is the observed dependent variable; w_i is the set of individual characteristics explaining participation decision; x_i is the set of variables explaining the expenditure or consumption decision of the household; ϵ_1, ϵ_2, and ε are the independent, homoscedastic, normally distributed error terms, respectively.

The tobit model assumes that participation and consumption decisions can be modelled as a single equation, whereas Cragg (1971) relaxed this assumption, and models both of these as separate regression equations introducing the double hurdle model.

Part II
Distributional Pattern

Chapter 4
Distributional Pattern of Fuel Consumption in India

4.1 Introduction

From the previous chapter, we observed that ten major alternative fuels are available for cooking purposes in India. These fuels are (i) coke or coal, (ii) firewood and chips, (iii) LPG, (iv) gobar gas, (v) dung cakes, (vi) charcoal, (vii) kerosene, (viii) electricity, (ix) others and (x) no cooking arrangements. Using Stata software, we have extracted data and prepared tables showing sample distribution among ten fuels within as well as across states of India. Data tabulation reflects that out of these ten types of cooking fuel, rural households tend to prefer firewood as the most preferred fuel, and urban households are inclined more towards LPG, i.e. they consider LPG as the most preferred fuel. This chapter provides distributional pattern of fuel consumption in India and more highlights on rural India.

4.2 Type of Fuel Consumption in India

The above said ten types of fuels are used or consumed in India. Now we report the distribution of fuel consumption in India. The distributional pattern of fuel consumption is observed within and between states in India. Table 4.1 presents cross distributional pattern of fuel consumption across states and territories of India including urban and rural areas (see Daripa & Dinda, 2020). Each state has three rows in Table 4.1, showing state name, state share and national share. The sample of state households is distributed among ten fuels which are presented in columns of state row. State share row and national share row display the distribution of fuel consumption at state level and national level, respectively. Columns 2–11 display fuel types (*coke or coal, firewood, LPG, gobar gas, dung cakes, charcoal, kerosene,*

This chapter is partly based on Daripa and Dinda (2020).

S. Daripa, S. Dinda, *Hurdles in Fuel Choice in Rural India*, SpringerBriefs in Geography,
https://doi.org/10.1007/978-3-031-71605-8_4

Table 4.1 Distributional pattern of fuel consumption in India

State	Coke and coal	Firewood and chips	LPG	Gobar gas	Dung cake	Charcoal	Kerosene	Electricity	Others	No cooking arrangements	Total
Jammu & Kashmir	1	1447	1660	0	34	7	123	90	2	20	3384
State share	0.03	42.76	49.05	0	1	0.21	3.63	2.66	0.06	0.59	100
National share	0.06	3.22	3.82	0	0.72	5	4.87	27.95	0.1	1.07	3.33
Himachal Pradesh	1	1216	740	0	5	2	38	9	0	29	2040
State share	0.05	59.61	36.27	0	0.25	0.1	1.86	0.44	0	1.42	100
National share	0.06	2.71	1.7	0	0.11	1.43	1.5	2.8	0	1.55	2.01
Punjab	1	579	1802	22	496	0	130	10	48	28	3116
State share	0.03	18.58	57.83	0.71	15.9	0	4.17	0.32	1.54	0.9	100
National share	0.06	1.29	4.15	14.97	10.5	0	5.15	3.11	2.42	1.5	3.07
Chandigarh	0	18	222	0	4	0	50	0	0	18	312
State share	0	5.77	71.15	0	1.28	0	16.03	0	0	5.77	100
National share	0	0.04	0.51	0	0.08	0	1.98	0	0	0.96	0.31
Uttaranchal	1	812	903	2	5	0	37	0	0	22	1782
State share	0.06	45.57	50.67	0.11	0.28	0	2.08	0	0	1.23	100
National share	0.06	1.81	2.08	1.36	0.11	0	1.46	0	0	1.18	1.75
Haryana	0	720	1343	0	415	2	23	4	65	17	2589
State share	0	27.81	51.87	0	16	0.08	0.89	0.15	2.51	0.66	100
National share	0	1.6	3.09	0	8.79	1.43	0.91	1.24	3.28	0.91	2.55
Delhi	1	25	769	0	0	1	16	3	43	87	945
State share	0.11	2.65	81.38	0	0	0.11	1.69	0.32	4.55	9.21	100
National share	0.06	0.06	1.77	0	0	0.71	0.63	0.93	2.17	4.66	0.93
Rajasthan	5	2562	1434	0	24	1	48	3	4	50	4131
State share	0.12	62.02	34.71	0	0.58	0.02	1.16	0.07	0.1	1.21	100

National share	0.32	5.71	3.3	0	0.51	0.71	1.9	0.93	0.2	2.68	4.06
Uttar Pradesh	36	3948	2477	0	2231	3	56	14	196	53	9014
State share	0.4	43.8	27.48	0	24.8	0.03	0.62	0.16	2.17	0.59	100
National share	2.29	8.79	5.7	0	47.2	2.14	2.22	4.35	9.89	2.84	8.87
Bihar	75	2019	1160	0	836	2	22	3	445	17	4579
State share	1.64	44.09	25.33	0	18.3	0.04	0.48	0.07	9.72	0.37	100
National share	4.77	4.5	2.67	0	17.7	1.43	0.87	0.93	22.45	0.91	4.5
Sikkim	1	167	554	1	0	0	11	0	0	34	768
State share	0.13	21.74	72.14	0.13	0	0	1.43	0	0	4.43	100
National share	0.06	0.37	1.27	0.68	0	0	0.44	0	0	1.82	0.76
Arunachal Pradesh	3	800	826	0	1	0	21	17	1	5	1674
State share	0.18	47.79	49.34	0	0.06	0	1.25	1.02	0.06	0.3	100
National share	0.19	1.78	1.9	0	0.02	0	0.83	5.28	0.05	0.27	1.65
Nagaland	0	300	723	0	0	0	1	0	0	0	1024
State share	0	29.3	70.61	0	0	0	0.1	0	0	0	100
National share	0	0.67	1.66	0	0	0	0.04	0	0	0	1.01
Manipur	3	1115	1353	0	2	71	7	3	0	6	2560
State share	0.12	43.55	52.85	0	0.08	2.77	0.27	0.12	0	0.23	100
National share	0.19	2.48	3.11	0	0.04	50.71	0.28	0.93	0	0.32	2.52
Mizoram	1	324	1197	0	0	1	12	1	0	0	1536
State share	0.07	21.09	77.93	0	0	0.07	0.78	0.07	0	0	100
National share	0.06	0.72	2.75	0	0	0.71	0.48	0.31	0	0	1.51
Tripura	0	1360	459	1	0	0	26	0	0	10	1856

(continued)

Table 4.1 (continued)

State	Coke and coal	Firewood and chips	LPG	Gobar gas	Dung cake	Charcoal	Kerosene	Electricity	Others	No cooking arrangements	Total
State share	0	73.28	24.73	0.05	0	0	1.4	0	0	0.54	100
National share	0	3.03	1.06	0.68	0	0	1.03	0	0	0.54	1.83
Meghalaya	0	850	309	0	2	16	30	39	0	14	1260
State share	0	67.46	24.52	0	0.16	1.27	2.38	3.1	0	1.11	100
National share	0	1.89	0.71	0	0.04	11.43	1.19	12.11	0	0.75	1.24
Assam	3	1994	1345	7	3	1	38	0	10	38	3439
State share	0.09	57.98	39.11	0.2	0.09	0.03	1.1	0	0.29	1.1	100
National share	0.19	4.44	3.09	4.76	0.06	0.71	1.5	0	0.5	2.04	3.38
West Bengal	656	2494	2055	3	201	0	217	4	534	148	6312
State share	10.39	39.51	32.56	0.05	3.18	0	3.44	0.06	8.46	2.34	100
National share	41.76	5.56	4.73	2.04	4.26	0	8.59	1.24	26.94	7.93	6.21
Jharkhand	596	1404	591	0	57	1	17	6	1	64	2737
State share	21.78	51.3	21.59	0	2.08	0.04	0.62	0.22	0.04	2.34	100
National share	37.94	3.13	1.36	0	1.21	0.71	0.67	1.86	0.05	3.43	2.69
Orissa	59	2914	672	6	51	17	48	25	154	80	4026
State share	1.47	72.38	16.69	0.15	1.27	0.42	1.19	0.62	3.83	1.99	100
National share	3.76	6.49	1.55	4.08	1.08	12.14	1.9	7.76	7.77	4.29	3.96
Chhattisgarh	67	1553	412	5	47	2	34	12	5	37	2174
State share	3.08	71.44	18.95	0.23	2.16	0.09	1.56	0.55	0.23	1.7	100
National share	4.26	3.46	0.95	3.4	1	1.43	1.35	3.73	0.25	1.98	2.14
Madhya Pradesh	28	2726	1537	21	281	0	76	5	1	40	4715
State share	0.59	57.82	32.6	0.45	5.96	0	1.61	0.11	0.02	0.85	100
National share	1.78	6.07	3.54	14.29	5.95	0	3.01	1.55	0.05	2.14	4.64
Gujarat	7	1493	1529	10	17	0	218	2	94	59	3429
State share	0.2	43.54	44.59	0.29	0.5	0	6.36	0.06	2.74	1.72	100

National share	0.45	3.33	3.52	6.8	0.36	0	8.63	0.62	4.74	3.16	3.37
Daman & Diu	0	26	79	0	0	0	18	0	0	5	128
State share	0	20.31	61.72	0	0	0	14.06	0	0	3.91	100
National share	0	0.06	0.18	0	0	0	0.71	0	0	0.27	0.13
D & N Haveli	0	79	88	0	0	0	17	0	0	6	190
State share	0	41.58	46.32	0	0	0	8.95	0	0	3.16	100
National share	0	0.18	0.2	0	0	0	0.67	0	0	0.32	0.19
Maharashtra	11	2535	4467	26	7	4	384	7	374	229	8044
State share	0.14	31.51	55.53	0.32	0.09	0.05	4.77	0.09	4.65	2.85	100
National share	0.7	5.65	10.28	17.69	0.15	2.86	15.2	2.17	18.87	12.27	7.91
Andhra Pradesh	12	2735	3833	6	3	9	108	20	2	168	6896
State share	0.17	39.66	55.58	0.09	0.04	0.13	1.57	0.29	0.03	2.44	100
National share	0.76	6.09	8.82	4.08	0.06	6.43	4.28	6.21	0.1	9	6.78
Karnataka	0	1983	1763	21	0	0	169	15	0	145	4096
State share	0	48.41	43.04	0.51	0	0	4.13	0.37	0	3.54	100
National share	0	4.42	4.06	14.29	0	0	6.69	4.66	0	7.77	4.03
Goa	0	36	382	2	0	0	22	0	0	5	447
State share	0	8.05	85.46	0.45	0	0	4.92	0	0	1.12	100
National share	0	0.08	0.88	1.36	0	0	0.87	0	0	0.27	0.44
Lakshadweep	0	89	56	0	0	0	22	14	0	10	191
State share	0	46.6	29.32	0	0	0	11.52	7.33	0	5.24	100
National Share	0	0.2	0.13	0	0	0	0.87	4.35	0	0.54	0.19
Kerala	3	2274	2025	14	0	0	16	9	2	119	4462
State share	0.07	50.96	45.38	0.31	0	0	0.36	0.2	0.04	2.67	100

(continued)

Table 4.1 (continued)

State	Coke and coal	Firewood and chips	LPG	Gobar gas	Dung cake	Charcoal	Kerosene	Electricity	Others	No cooking arrangements	Total
National share	0.19	5.07	4.66	9.52	0	0	0.63	2.8	0.1	6.38	4.39
Tamil Nadu	0	2098	3962	0	0	0	343	6	0	237	6646
State share	0	31.57	59.61	0	0	0	5.16	0.09	0	3.57	100
National share	0	4.67	9.11	0	0	0	13.58	1.86	0	12.7	6.54
Pondicherry	0	83	426	0	1	0	18	0	0	48	576
State share	0	14.41	73.96	0	0.17	0	3.13	0	0	8.33	100
National share	0	0.18	0.98	0	0.02	0	0.71	0	0	2.57	0.57
A & N Islands	0	115	321	0	0	0	110	1	1	18	566
State share	0	20.32	56.71	0	0	0	19.43	0.18	0.18	3.18	100
National share	0	0.26	0.74	0	0	0	4.35	0.31	0.05	0.96	0.56
India	1571	44,893	43,474	147	4723	140	2526	322	1982	1866	1,01,644
India share	1.55	44.17	42.77	0.14	4.65	0.14	2.49	0.32	1.95	1.84	100
National share	100	100	100	100	100	100	100	100	100	100	100

Source: Daripa and Dinda (2020)

electricity, others and no cooking arrangements) and the last column is row total. Table 4.1 describes the distributional pattern of household fuel consumption in aggregate for each state and union territory in India. Table 4.1 shows the distribution of sample households as well as shares of fuel consumption between fuel types and across states in India.

Table 4.1 shows each fuel consumption of Indian states and union territories. Fuel consumption of each state of India is shown in the row, while the total consumption of each fuel type is displayed in the columns. Along the rows, we observe the share of each fuel in the total fuel consumption of the particular state, while along the columns we observe the share of each state in total consumption of specific fuel type in India. Daripa and Dinda (2020) have described the fuel consumption pattern of all states and union territories in detail; however, here, we discuss some selected states and their fuel consumption pattern for this study purpose.

From Table 4.1, it is observed that Jammu & Kashmir (J&K) consumes LPG around 49.05% of total state fuel consumption followed by firewood and chips (42.76%). Kerosene (3.63%) and electricity (2.66%) are alternative fuels that are also used in J&K. Two main fuels consumed in Jammu & Kashmir are LPG and firewood and chips. The national consumption shares of LPG and firewood and chips are 3.82% and 3.22%, respectively. In terms of *national share, Jammu & Kashmir has the highest (27.95%) electricity consumption share in India.* Similarly, Table 4.1 displays that 57.83% of households in Punjab use LPG followed by firewood and chips (18.58%) and dung cake (15.92%). However, national shares of this consumption are 1.29% (for firewood and chips), 4.15% (for LPG) and 10.5% (for dung cake). In Chandigarh, 71.15% of households consume LPG followed by *kerosene (16.03%)* and firewood and chips (5.77%). In Haryana, 51.87% of households consume LPG followed by firewood and chips (27.81%) and *dung cake* (16.03%). In Delhi, 81.38%, 2.65% and 1.69% of households consume LPG, firewood and chips, and kerosene, respectively. *No cooking arrangements is found in 9.21%* households in Delhi.

In Uttar Pradesh (UP), 43.8%, 27.48% and 24.75% of households consume, respectively, firewood and chips, LPG and *dung cakes*. The national consumption share of these fuels are 8.79% for firewood and chips, 5.7% for LPG and 47.24% for *dung cakes*. Among these, the national share of firewood and chips, and dung cakes are the highest in UP. In Bihar, 44.09%, 25.33% and *18.26%* of households consume, respectively, firewood and chips, LPG and dung cakes. In Bihar, the national share of firewood and chips, and dung cake are 4.5% and 17.7%, respectively.

In West Bengal, 39.51%, 32.56% and *10.39%* of households consume, respectively, firewood and chips, LPG and *coke or coal*. The respective national shares of these fuels are 5.56%, 4.73% and 41.76%. Among these, the *national share of coke or coal consumption of West Bengal is the highest in India.*

In Maharashtra, 55.53%, 31.51% and 4.77% of households consume LPG, firewood and chips, and kerosene, respectively; however, their respective shares in India are 10.28%, 5.65% and 15.2%. Among these, national *shares of LPG and kerosene are the highest in India*. In Kerala, 50.96% and 45.38% of households consume, respectively, firewood and chips, and LPG, but their respective national shares are 5.07% and 4.66%. In Tamil Nadu (TN), 59.61% and 31.57% of households use, respectively, LPG, and firewood and chips, but their respective national shares are 9.11% and 4.67%. In a similar way, we may discuss other states and union territories of India.

Table 4.1 indicates that the state having the highest share of LPG consumption in India is Maharashtra (10.28%) followed by Tamil Nadu (9.11%) and Andhra Pradesh (8.82%). Uttar Pradesh has the highest national share of firewood and chips consumption (8.79%) followed by Orissa (6.49%) and Andhra Pradesh (6.09%) in India. West Bengal (41.76%) has the highest national share of coke or coal consumption in India followed by Jharkhand (37.94%) and Bihar (4.77%). Maharashtra (17.69%), Punjab (14.97%) and Karnataka (14.29%) have the highest national shares of gobar gas consumption, while Uttar Pradesh (47.24%) has the highest national share of dung cake, followed by Bihar (17.7%) and Punjab (10.5%). Manipur, Orissa and Meghalaya lead in charcoal consumption, while Maharashtra, Tamil Nadu and Gujarat lead in kerosene consumption. It is also observed that Jammu & Kashmir, Meghalaya and Orissa have the highest share in India's electricity consumption.

4.3 Distribution of Fuel Consumers in Rural and Urban India

From the data, it is observed that in India, individuals have ten major alternatives for cooking purposes, namely coke or coal, firewood, LPG, gobar gas, dung cakes, charcoal, kerosene, electricity, others and no cooking arrangements. Out of these ten types of cooking fuels, rural households tend to prefer firewood and chips as the most preferred fuel, while urban households are inclined more towards LPG, i.e. urban households consider LPG as the most preferred fuel. From Table 3.1, we have also observed that firewood and chips is the most preferred fuel in rural India and LPG in urban India. Now, we attempt to identify state-wise fuel preference patterns. For analytical purposes, we have ranked fuel preferences in terms of the most preferred fuel of each state in urban and rural India. We also observe the consumption pattern of different fuels in rural as well as in urban India; however, results are different.

Considering fuel consumption pattern in rural India, it is observed that firewood is the most preferred fuel in 25 (23 states and 2 union territories) states, while LPG is the most preferred fuel in ten states (5 states and 5 union territories). However, considering fuel consumption pattern in urban India, it is noted that LPG is the most preferred fuel in all states and union territories, while firewood and chips is the most preferred fuel in no state. These results are reflected in Fig. 4.1.

So, clean fuel such as LPG is the most preferred fuel in urban India, while dirty fuel like firewood and chips dominates in rural India. Hence, it can be inferred that

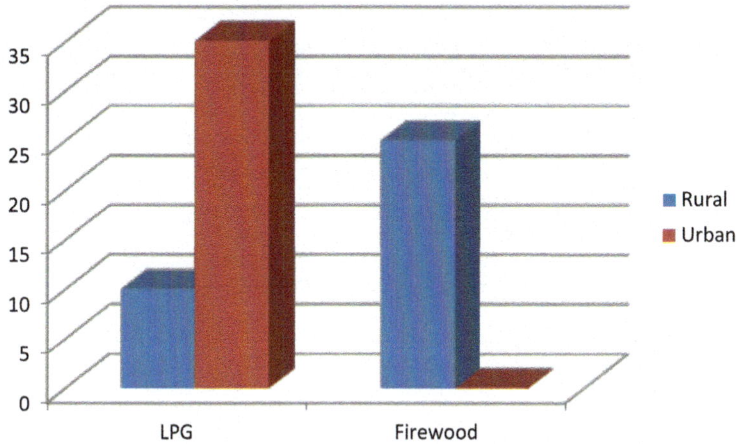

Source: Daripa and Dinda (2020)

Fig. 4.1 State-wise most preferred fuel of rural and urban India in 2011–12. (Source: Daripa & Dinda, 2020)

firewood and chips is the most preferred fuel in rural India and LPG in urban India among the available ten sources.

It is also observed that Uttar Pradesh, Orissa and Andhra Pradesh have the highest shares of firewood and chips consumption in rural India.

4.4 Cooking Fuel-Wise Income Distribution in Rural India

This study focuses on the choice of firewood and chips in rural India and identifies its determining factors which are considered one by one. The most important determinant of fuel choice of households is income, which is not available. So, following the literature, we consider monthly per capita consumption expenditure (MPCE) as a proxy variable for income. Now we observe the income distribution of households in rural India corresponding to primary source-wise cooking fuels. Table 4.2 displays the percentage distribution of households in the fractal class of MPCE with primary source of cooking fuels in rural India.

Table 4.2 shows the percentage distribution of rural households with their average MPCE corresponding to each primary source of cooking fuel in India. Our study also focuses on the fact that firewood and chips consumed by rural households comes from the major ten sources as shown in Table 4.2. It is observed that households' firewood and chips consumption declines, while that of LPG increases with higher-income fractal classes.

Table 4.2 Percentage distribution of households in fractal class of income with primary source of cooking fuels in rural India

Fractal class of income (MPCE)	Primary source of cooking fuels in rural India									
	Coke and coal	Firewood and chips	LPG	Gobar gas	Dung cake	Charcoal	Kerosene	Electricity	Others	No cooking arrangement
0–5%	0.4	81.0	0.2	0	10.9	0	0.7	0	5.8	1.0
5–10%	1.4	77.4	0.7	0	13.0	0	0.1	0	7.2	0.1
10–20%	1.5	76.1	1.5	0	13.2	0	0.2	0	6.9	0.6
20–30%	1.5	77.7	1.9	0.2	11.2	0	0.3	0	6.8	0.3
30–40%	1.5	75.3	4.5	0	12.0	0	0.5	0	6.0	0.1
40–50%	1.4	74.5	6.9	0.1	10.3	0	0.8	0	5.2	0.8
50–60%	1.0	72.8	9.8	0.1	9.4	0	0.5	0	5.7	0.7
60–70%	1.2	69.6	14.3	0.2	8.4	0	1.1	0	4.3	0.9
70–80%	1.0	65.7	19.0	0.2	8.3	0	1.3	0.1	3.5	1.0
80–90%	1.1	59.4	26.2	0.3	7.7	0	1.3	0	2.8	1.0
90–95%	0.8	45.5	39.8	0.5	8.8	0	1.2	0.1	1.5	1.7
95–100%	0.3	33.7	49.2	0.7	5.1	0.1	1.7	0.4	1.0	7.8
Average	1.1	67.3	15.0	2.0	9.6	0	0.9	0.1	4.6	1.3
Average MPCE (₹)	1224.7	1288.2	2277.5	2465.3	1275.1	2502.0	1640.8	2718.5	1152.1	3437.1

Source: 68th Round NSSO Report

4.5 Distribution of Fuel Choices of Rural Household Across States of India

This study is mainly focused on fuel selection and their consumption in rural India. We have used National Sample Survey Office (NSSO) 68th round data to explain the fuel choice patterns of households of rural area in India. We consider a brief overview about the nature of the data. From the data, it is observed that in India, individuals are faced with ten major alternative sources of fuels for their energy consumption purposes, namely coke or coal, firewood, LPG, gobar gas, dung cakes, charcoal, kerosene, electricity, others and no cooking arrangements. Our preliminary analysis begins with state-wise choice of major fuels in India and attempts to identify fuel choice preferences of each state. We have performed a state-wise tabulation of the major fuels consumed in rural India (Table 4.3).

Table 4.3 represents the consumption of different fuels in rural India. The row totals reflect the total consumption of all fuels in rural sector of each state, while the column totals reflect the total consumption of each fuel in India's rural sector as a whole. Along the row, we observe the share of each fuel in total fuel consumption of rural sector of each state. Along the columns, the share of each state is observed in the total consumption of a particular fuel in rural India.

Table 4.3 provides energy consumption shares of all fuels in rural sector across states and union territories. For example, Table 4.3 shows that 30.46% of the total fuel consumption of rural parts of Jammu & Kashmir (J&K) comes from LPG which is followed by firewood and chips that comprises 63.63% of the total fuel consumption of the state. Other important fuels are kerosene consumed by 3.05% of households and electricity by 0.84% of households in rural J&K. Table 4.3 clearly displays that LPG and firewood are the two main fuels consumed in rural parts of Jammu & Kashmir. Similarly, we can discuss it for rural parts of all states and territories. Here we summarise all in briefly.

The results of Table 4.3 show that the state having the highest share of firewood and chips among the available ten fuels within the said state is Chhattisgarh (89.51%) followed by Meghalaya (88.67%) and Tripura (88.03%). Similarly, it also shows that the state having the highest share of LPG consumption is Delhi (93.65%) followed by Chandigarh (79.69%) and Goa (75.47%). Jharkhand (17.02%) has the highest state share in coke and coal consumption followed by West Bengal (7.54%). It is observed that the state consumption share of dung cake is the highest in Uttar Pradesh (33.54%) followed by Punjab (27.58%), Haryana (26.42%) and Bihar (22.84%).

Table 4.3 State-wise distribution of fuel choices of rural households across India

State	Coke and coal	Firewood and chips	LPG	Gobar gas	Dung cake	Charcoal	Kerosene	Electricity	Others	No cooking arrangements	Total
J&K	0	63.63	30.46	0	1.57	0.1	3.05	0.84	0.1	0.25	100
HP	0.06	70.85	27.52	0	0.06	0	0.91	0.3	0	0.3	100
Punjab	0	29.25	36.73	1.29	27.6	0	1.68	0.19	2.9	0.39	100
Chandigarh	0	1.56	79.69	0	6.25	0	12.5	0	0	0	100
Uttaranchal	0.1	69.85	28.72	0.19	0.1	0	0.67	0	0	0.38	100
Haryana	0	44.06	24.31	0	26.4	0	0.63	0.21	4.29	0.07	100
Delhi	0	3.17	93.65	0	0	0	0	0	0	3.17	100
Rajasthan	0	85.96	12.33	0	0.74	0.04	0.5	0.12	0.16	0.16	100
UP	0.15	53.31	9.6	0	33.5	0	0.1	0.02	3.01	0.27	100
Bihar	0.6	50.79	12.9	0	22.8	0	0.42	0.03	12.3	0.12	100
Sikkim	0.16	26.97	69.41	0.16	0	0	0.49	0	0	2.8	100
Arunachal Pradesh	0.19	67.92	29.83	0	0.09	0	1.22	0.19	0.09	0.47	100
Nagaland	0	33.04	66.96	0	0	0	0	0	0	0	100
Manipur	0.15	57.63	41.21	0	0.07	0.58	0.07	0	0	0.29	100
Mizoram	0	43.91	55.16	0	0	0	0.94	0	0	0	100
Tripura	0	88.03	11.28	0.08	0	0	0.3	0	0	0.3	100
Meghalaya	0	88.67	8.88	0	0	0	0.58	1.52	0	0.35	100
Assam	0.04	70.73	28.35	0.27	0	0.04	0.35	0	0.08	0.15	100
WB	7.54	60.24	12.28	0.08	5.19	0	0.45	0	13.97	0.25	100
Jharkhand	17.02	73.48	5.18	0	3.02	0	0.23	0	0.06	1.02	100
Orissa	0.87	83.86	7.53	0.2	1.58	0.13	0.27	0.34	4.84	0.37	100
Chhattisgarh	0.69	89.51	5.21	0.28	2.64	0	0.9	0.21	0.07	0.49	100
MP	0.18	79.31	9.69	0.77	9.25	0	0.44	0.04	0	0.33	100
Gujarat	0.06	72.84	20.56	0.53	0.76	0	4.61	0.12	0.06	0.47	100

Daman & Diu	0	29.69	46.88	0	0	0	21.88	0	0	1.56	100
D & N Haveli	0	76.04	13.54	0	0	0	6.25	0	0	4.17	100
Maharashtra	0.02	53.91	34.61	0.65	0.17	0.07	1.56	0	7.99	1.02	100
Andhra Pradesh	0.18	58.75	39.49	0.15	0.08	0.15	0.38	0.03	0.05	0.74	100
Karnataka	0	74.8	21.34	0.93	0	0	1.86	0	0	1.07	100
Goa	0	16.35	75.5	1.26	0	0	6.92	0	0	0	100
Lakshadweep	0	82.54	4.76	0	0	0	4.76	6.35	0	1.59	100
Kerala	0.08	59.7	37.96	0.54	0	0	0.19	0.23	0	1.3	100
Tamil Nadu	0	48.78	47.7	0	0	0	2.53	0.03	0	0.96	100
Pondicherry	0	24.22	68.75	0	0.78	0	3.13	0	0	3.13	100
A & N Islands	0	38.85	43.17	0	0	0	15.47	0.36	0	2.16	100
India	1.1	62.68	24.4	0.24	7.04	0.04	1.04	0.13	2.8	0.54	100

Source: Author's calculation based on NSSO 68th round data

4.6 State-Wise Most Preferred Fuel in Rural India

Earlier discussion and results suggest that firewood and chips is the most preferred fuel in 25 (23 states and 2 union territories) states in rural India, while LPG is the most preferred fuel in ten states (5 states and 5 union territories). This result is reflected in Fig. 4.2.

Hence, it can be inferred that the most preferred fuel in rural India is firewood and chips among available sources. So, the main focus would be on the choice of firewood and chips consumption in rural India. It is also observed that Uttar Pradesh, Orissa and Andhra Pradesh have the highest national shares in firewood and chips consumption in rural India. Shares of each state in rural India's firewood and chips consumption are represented later.

It is also reflected from Table 4.1 that firewood is the second most preferred fuel of 18 major states of India; LPG is the second most preferred fuel for 14 major states of India (Fig. 4.3). It should be noted that coke and coal, kerosene and no cooking arrangement are also turned to be the second most preferred for one major state of India for each fuel (Daripa & Dinda, 2020).

Table 4.1 also shows the third most preferred fuel of different states of India. The results are reflected from Fig. 4.4 which shows that kerosene is the third most preferred fuel of 16 major states of India followed by dung cake and no cooking arrangements in five major states and remaining fuels in the other states.

This choice of fuel is, however, governed by a number of factors. We will consider them one by one in the next part of our discussion. Among these, income is the most important determinant of fuel choice of households. The data on income is, however, not available. So, we consider a proxy variable for income which is monthly per capita consumption expenditure (MPCE). We have computed the weighted mean MPCE for the rural sector of each state as well as for the rural sector of the nation as a whole and discuss it in the following chapters.

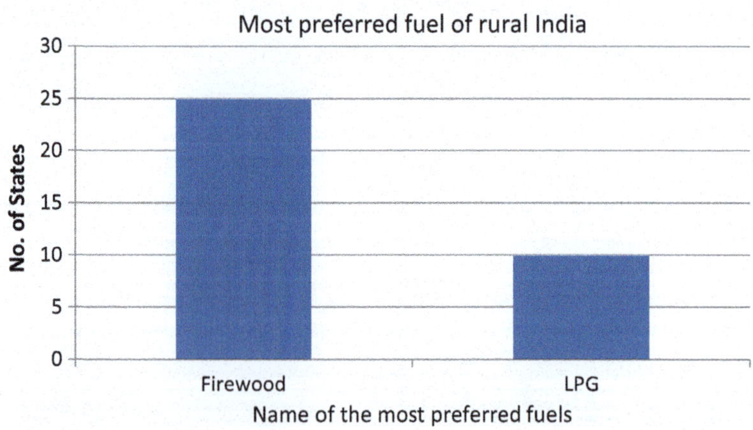

Source: Daripa and Dinda (2020)

Fig. 4.2 State-wise most preferred fuel in rural India. (Source: Daripa & Dinda, 2020).

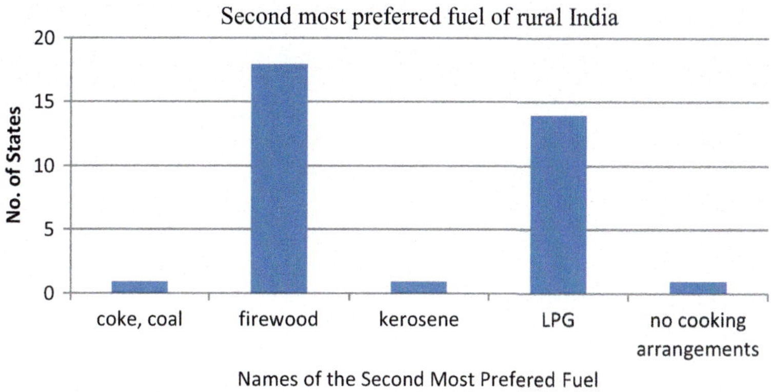

Source: Daripa and Dinda (2020)

Fig. 4.3 Second most preferred fuels in India. (Source: Daripa & Dinda, 2020)

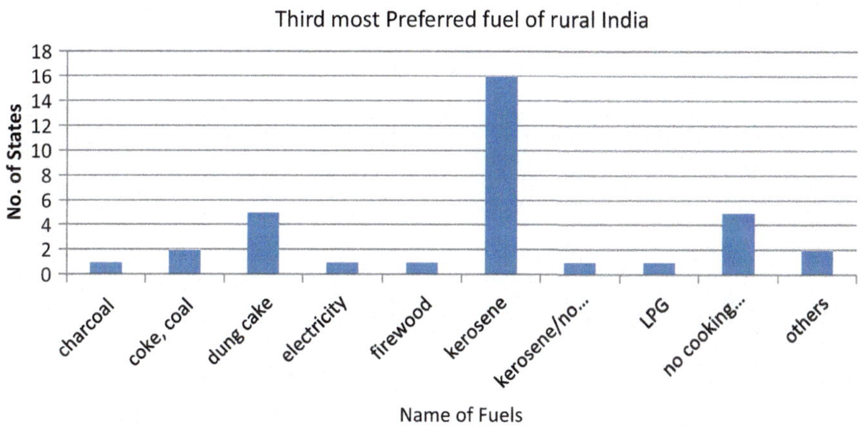

Source: Daripa and Dinda (2020)

Fig. 4.4 Third most preferred fuel in India. (Source: Daripa & Dinda, 2020)

Chapter 5
Distributional Pattern of Firewood and Chips Consumption in Rural India

5.1 Shares of Firewood and Chips Consumption in Rural Area Across States in India

From the previous chapter, it is clear that clean fuel LPG is the most preferred fuel in urban India, while dirty fuels like firewood and chips dominates in rural India. Hence, it can be inferred that firewood is the most preferred fuel in rural India among the available ten sources. So, our main focus is on the choice of firewood consumption in rural India. It is also observed that Uttar Pradesh, Orissa and Andhra Pradesh have the highest shares in firewood consumption in rural India. Shares of each state in rural India's firewood consumption are represented in Table 5.1.

Table 5.1 reflects that Uttar Pradesh has the highest share in rural India's firewood and chips consumption followed by Orissa and Andhra Pradesh. The top 18 states out of 35 (28 states and 7 territories) which have the highest shares in rural India's total firewood and chips consumption are considered high-firewood-consuming states in India and they are coloured red in Table 5.1. The remaining states are considered low-firewood-consuming states in our preliminary analysis. The results reflecting the share of each state in rural India's firewood consumption are represented using Fig. 5.1 as follows:

5.2 Source-Wise Distribution of Firewood and Chips in Rural India

Table 5.2 describes the percentage of household consuming firewood and chips of each source and their corresponding income distribution. Table 5.2 suggests that a majority of firewood and chips consumers rely on free collection of firewood, i.e. 34.32% firewood and chips consumers rely on free collection. It is also clearly

S. Daripa, S. Dinda, *Hurdles in Fuel Choice in Rural India*, SpringerBriefs in Geography,
https://doi.org/10.1007/978-3-031-71605-8_5

Table 5.1 National shares of firewood and chips consumption in rural areas across states in India

State	Firewood	State	Firewood	State	Firewood
Uttar Pradesh	8.43	Jammu & Kashmir	3.46	Mizoram	0.75
Orissa	6.67	Jharkhand	3.45	Nagaland	0.59
Andhra Pradesh	6.16	Chhattisgarh	3.45	Sikkim	0.44
Rajasthan	5.93	Gujarat	3.33	A & N Islands	0.29
Maharashtra	5.81	Himachal Pradesh	3.14	D & N Haveli	0.2
Madhya Pradesh	5.8	Tripura	3.09	Lakshadweep	0.14
West Bengal	5.74	Manipur	2.12	Pondicherry	0.08
Assam	4.93	Meghalaya	2.03	Goa	0.07
Bihar	4.49	Uttaranchal	1.96	Daman & Diu	0.05
Tamil Nadu	4.33	Arunachal Pradesh	1.94	Delhi	0.01
Kerala	4.16	Haryana	1.68	Chandigarh	0
Karnataka	4.1	Punjab	1.21		

Source: Author

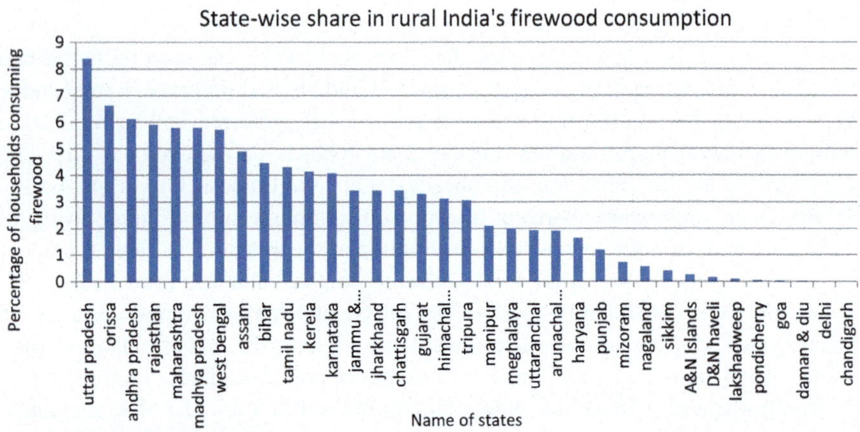

Source: Daripa and Dinda (2020)

Fig. 5.1 State-wise national share in rural India's firewood consumption (Source: Daripa & Dinda, 2020)

observed that 30.05% and 27.54% of firewood and chips consumers rely on only purchase and home grown of firewood and chips, respectively. Table 5.2 describes the summary statistics of MPCE across each of the above groups of firewood and chips consumers in rural India. It is noted that the mean of MPCE of a free collection of firewood and chips group is lower than that of only purchase and home-grown firewood and chips consumers.

Table 5.2 Firewood source-wise distribution of households and corresponding mean income in Rural India

Sources of firewood	Percentage freq.	Mean MPCE	Standard deviation MPCE	Coefficient of variation MPCE
Only purchase	30.05	1669.24	1165.86	0.698
Only home-grown stock	27.54	2009.01	2328.28	1.159
Both purchase and home grown	3.26	1711.41	2292.64	1.34
Only free collection	**34.32**	1465.71	946.31	0.646
Exchange of goods and services	0.19	1342.99	731.36	0.545
Gifts/charities	0.18	1785.55	992.73	0.556
Others	4.46	1429.14	862.55	0.604
Total	100			

Source: Daripa and Dinda (2022)

5.3 Firewood and Chips Consumption Pattern of Poor and Rich States in India

We observe that some states lie above the national average MPCE, while others lie below it. States like Uttar Pradesh, Bihar, Manipur, Tripura, Meghalaya, Assam, West Bengal, Jharkhand, Orissa, Chhattisgarh and Madhya Pradesh lie below the national average MPCE and these states are termed as poor, while states such as Jammu & Kashmir, Himachal Pradesh, Punjab, Uttaranchal, Haryana, Rajasthan, Sikkim, Arunachal Pradesh, Nagaland, Mizoram, Karnataka, Gujarat, Maharashtra, Andhra Pradesh, Goa, Kerala and Tamil Nadu are non-poor or rich as these states lie above the national average MPCE.

Again, we also observe that high-firewood and chips–consuming states are Uttar Pradesh, Orissa, Andhra Pradesh, Rajasthan, Maharashtra, Madhya Pradesh, West Bengal, Assam, Bihar, Tamil Nadu, Kerala, Karnataka, Jammu & Kashmir, Jharkhand, Chhattisgarh, Gujarat, Himachal Pradesh and Tripura, while low-firewood and chips–consuming states are Manipur, Meghalaya, Uttaranchal, Arunachal Pradesh, Haryana, Punjab, Nagaland, Mizoram, Sikkim, D & N Haveli, A & N Islands, Lakshadweep, Pondicherry, Goa, Daman & Diu, Delhi and Chandigarh. Comparing the poor and non-poor states with high and low-firewood and chips–consuming states, we make a comparative analysis of state-wise fuel consumption patterns in rural India. The results are represented in Table 5.3.

It is also observed that Uttar Pradesh, Orissa, Madhya Pradesh, West Bengal, Assam, Bihar, Jharkhand, Chhattisgarh and Tripura are rural poor in terms of income but high-firewood and chips–consuming states. This indicates that rural poor states use more dirty fuels like high firewood and chips. However, being rural poor, Manipur and Meghalaya consume low firewood and chips.

Table 5.3 Firewood and chips consumption pattern of rural rich and poor states in India

	High firewood and chips consuming states *(greater than national average)*	Low firewood and chips consuming states *(lower than national average)*
Rural rich *(more than the national average MPCE)*	Andhra Pradesh, Rajasthan, Maharashtra, Tamil Nadu, Kerala, Karnataka, Jammu & Kashmir, Gujarat, Himachal Pradesh	Uttaranchal, Arunachal Pradesh, Haryana, Punjab, Mizoram, Nagaland, Sikkim, A & N Islands, Lakshadweep, Pondicherry, Goa, Daman & Diu, Delhi, Chandigarh
Rural poor *(less than national average MPCE)*	Uttar Pradesh, Orissa, Madhya Pradesh, West Bengal, Assam, Bihar, Jharkhand, Chhattisgarh, Tripura	Manipur, Meghalaya, D & N Haveli

Source: Author

Being rural rich, states like Uttaranchal, Arunachal Pradesh, Haryana, Punjab, Mizoram, Nagaland, Sikkim, A & N Islands, Lakshadweep, Pondicherry, Goa, Daman & Diu, Delhi and Chandigarh use low firewood and chips. Comparing these results, we find that mostly rural rich states consume less firewood and chips, whereas rural poor states consume it more.

5.4 Land Distributional Pattern Across States in Rural India

Apart from income, the major asset of rural households is considered as land holding. We have also considered the land-holding pattern of households focusing on land owned by households. It is seen that we have data on land holdings for only 25,794 households out of total 59,683 rural households. This implies that 33,889 households are landless. The mean land holding of 25,794 households is 0.0102 hectares and the standard deviation of land holding is 0.0029 hectares.

We have also considered the land-holding pattern of households in the determination of fuel choice focusing on land owned, land possessed and land cultivated. Since our focus is on rural areas and choice of firewood and chips, we compare the state-wise mean land holdings of owned, possessed and cultivated lands with high-firewood and chips–consuming states. It is observed that there are some land-rich states which are high consumers of firewood and chips like West Bengal, Bihar, Jharkhand, Tripura, Uttar Pradesh and Rajasthan. Of these states, all are poor states according to our estimate except Rajasthan. Also, there are some land-poor states that are high consumers of firewood and chips like Andhra Pradesh, Maharashtra, Karnataka, Tamil Nadu and Gujarat, but they all belong to the category of non-poor in our estimates. Apart from this, other high-firewood and chips–consuming states are either rich in owned land or possessed land or cultivated land or any two of them as presented in Table 5.4.

Table 5.4 Land distribution pattern across states in rural India

Land owned (LO)		Land possessed (LP)		Land cultivated (LC)	
Above national average	Below national average	Above national average	Below national average	Above national average	Below national average
Punjab	Jammu & Kashmir	Jammu & Kashmir	Punjab	Himachal Pradesh	Jammu & Kashmir
Rajasthan	Himachal Pradesh	Himachal Pradesh	Chandigarh	Uttaranchal	Punjab
Uttar Pradesh	Chandigarh	Uttaranchal	Haryana	Rajasthan	Chandigarh
Bihar	Uttaranchal	Rajasthan	Delhi	Uttar Pradesh	Haryana
Nagaland	Haryana	Uttar Pradesh	Sikkim	Bihar	Delhi
Manipur	Delhi	Bihar	Arunachal Pradesh	Tripura	Sikkim
Mizoram	Sikkim	Manipur	Nagaland	Meghalaya	Arunachal Pradesh
Tripura	Arunachal Pradesh	Tripura	Mizoram	Assam	Nagaland
Meghalaya	Assam	Meghalaya	Madhya Pradesh	West Bengal	Manipur
West Bengal	Chhattisgarh	Assam	Gujarat	Jharkhand	Mizoram
Jharkhand	Gujarat	West Bengal	Daman & Diu	Lakshadweep	Orissa
Orissa	Daman & Diu	Jharkhand	D & N Haveli		Chhattisgarh
Madhya Pradesh	D & N Haveli	Orissa	Maharashtra		Madhya Pradesh
Lakshadweep	Maharashtra	Chhattisgarh	Andhra Pradesh		Gujarat
Kerala	Andhra Pradesh	Lakshadweep	Karnataka		Daman & Diu
	Karnataka	Kerala	Goa		D & N Haveli
	Goa	A & N Islands	Tamil Nadu		Maharashtra
	Tamil Nadu		Pondicherry		Andhra Pradesh
	Pondicherry				Karnataka
	A & N Islands				Goa
					Kerala
					Tamil Nadu
					Pondicherry
					A & N Islands

Source: Author

5.5 Sources of Firewood and Chips Consumption and Land Percentile

We shall now discuss the sources of firewood and chips. The National Sample Survey Office (NSSO) classifies the sources of firewood and chips into six broad categories, namely (i) purchase, (ii) home-grown stock, (iii) both purchase and home-grown, (iv) free collection, (v) exchange of goods and services and (vi) gifts and charities or some others unknown sources. From descriptive statistics of sources of cooking fuel, it may be readily observed that the frequency of free collectors is the highest (Table 3.2). Besides free collection, most of the households either purchase this item or consume it from home-grown products. It is readily observable that the frequency is very small for other categories.

In order to have an idea of the standard of living of the households based on these categories, we have also presented the average MPCE and land holdings of these households. It is clearly visible that the non-consumers of firewood are the richest sub-groups, in terms of MPCE. However, these households have very small land holdings, implying the possibility that very few of them are engaged in agricultural activities. The average land holdings of the households belonging to the home-grown categories are much higher compared to the rest. In fact, the average MPCE is also higher, compared to others, excluding the non-consumers. It is likely that these households mostly belong to the class of rich farmers. The free collectors although have a very low MPCE, but land holdings are decent. Perhaps these households mostly belong to the class of poorer and marginal farmers.

For further explorations, in Fig. 5.2, we plot the amount of firewood and chips consumption for three main sources on different income and land quantiles. Note that if we focus on the total consumption, the shape is almost similar to an inverted-U. This obviously implies as income (amount of land holdings) increases, total consumption also increases but declines at very rich quantiles. Consumption, following the home-grown production, steadily increases with MPCE, and the increment is more in higher quantiles. The same category also exhibits a similar pattern moving to the higher quantiles of land. In fact, there is a small decline at the bottom and/or lower quantiles, but in the case of higher quantiles, the curve becomes much steeper. Free collection remains steady up to the eighth percentile of MPCE but starts to decline afterwards, whereas free collection slowly rises with land percentile up to fourth percentile and suddenly falls after it and remains stable for the rest of the higher land percentiles.

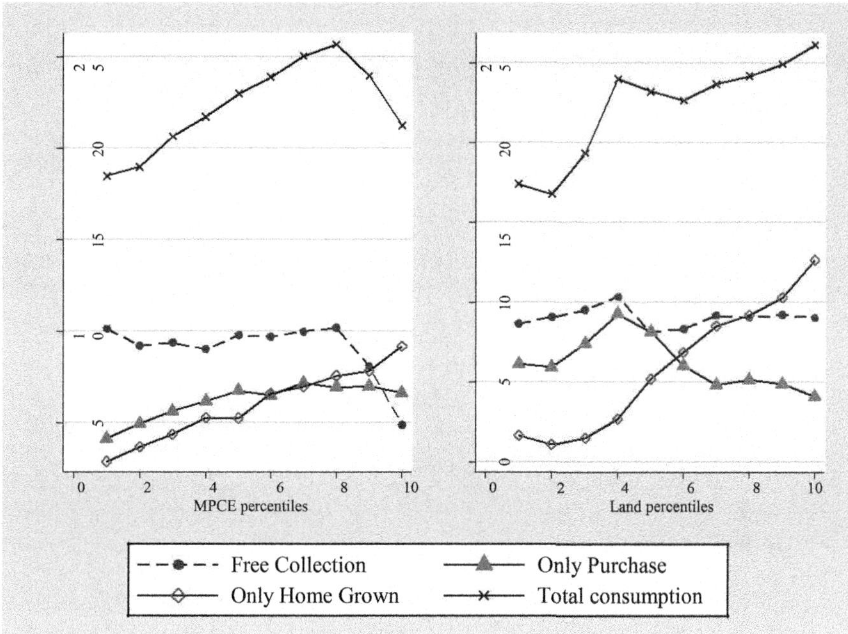

Source: Author

Fig. 5.2 Source-wise firewood and chips consumption pattern at MPCE and land percentiles. (Source: Author)

5.6 Conclusion

This chapter provides the distributional patterns of firewood and chips consumption share in rural areas across states of India, and it also highlights rich and poor states. This chapter provides a comparison of the source-wise firewood and chips consumption patterns over percentiles of MPCE and land. All percentile rural households raise firewood and chips consumption for home-grown but rich percentiles reduce free collections. Land-rich rural households reduce purchase only.

Part III
Determinants

Part II
Determinants

Chapter 6
Fuel Consumption in Rural India

6.1 Introduction

This study focuses on rural India and its fuel consumption. Indian rural households mostly use firewood and chips as their cooking fuel out of ten alternative fuels. In this context, this chapter attempts to answer the question of determining factors of fuel consumption in rural India. Following literature, traditional limited dependent models are used for estimation purposes in this chapter, and results are reported in two sections. Section 6.1 provides the results of determining factors of alternative fuels, while Sect. 6.2 reports only firewood and chips and also its major sources.

6.2 Section A: Alternative Fuels

The type of fuel consumption patterns and their determinants are estimated focusing on the consumption of selected major fuels of rural India. It was observed from the data that the major cooking fuel in rural India is firewood and chips, as it is a dirty fuel, so our analysis is mainly centred on estimating the hurdles in firewood and chips consumption patterns in rural India. However, we also estimate the above-said model for other fuels like LPG, dung cake, coke and coal, etc. Initially, we apply the probit model for the selection of the type of fuel on the basis of binary data, and next we use the truncated regression (or tobit) model for estimating the amount of consumption based on positive response only. The results of the two hurdles of each fuel consumption in rural India are reported in Table 6.1 as follows.

Table 6.1 provides the estimated results of the probit and truncated regression models. The probit model estimates the determinants of the probability of fuel selection of households in rural India, while the truncated regression model estimates the determinants of the amount of selected fuel consumption. The results of

S. Daripa, S. Dinda, *Hurdles in Fuel Choice in Rural India*, SpringerBriefs in Geography,
https://doi.org/10.1007/978-3-031-71605-8_6

Table 6.1 Probability of selection of different fuels and consumption in rural India

	LPG		Coke and coal		Dung cake	
Variable name	Probit	Truncated regression	Probit	Truncated regression	Probit	Truncated regression
Constant	−32.8*** (−37.52)	−3.569*** (−7.87)	−13.86*** (−4.54)	−2.397 (−0.40)	1.381 (1.32)	
Log_MPCE	7.104*** (31.68)	1.083*** (9.54)	3.588*** (4.24)	1.307 (0.79)	−0.574** (−2.04)	
LogMPCE_sq	−0.401*** (−27.82)	−0.0585*** (−8.23)	−0.276*** (−4.7)	−0.091 (−0.81)	0.029 (1.53)	
LO	22.0*** (17.36)	8.686*** (13.17)	13.15*** (4.56)	2.824 (0.43)	−0.76 (−0.47)	
Hhsize	0.028*** (7.04)	0.0563*** (26.45)	−0.042*** (−4.08)	0.037 (1.84)	0.069*** (15.40)	
Age	−0.013*** (−4.21)	−0.0032 (−1.89)	0.021*** (2.82)	0.0303 (1.89)	−0.017*** (−4.43)	
Age_sq	0.0001*** (3.52)	0.000004 (0.20)	−0.0003*** (−3.15)	−0.0004** (−2.01)	0.0002*** (3.22)	
Edu	0.1117*** (22.73)	0.0258*** (10.02)	−0.0006 (−0.05)	−0.0233 (−0.86)	−0.079*** (−12.02)	
Age_head	0.0215*** (6.27)	0.0136*** (7.28)	−0.0018 (−0.22)	0.0231 (1.40)	0.0086* (1.94)	
Age_head_sq	−0.0001*** (−3.91)	−0.0001*** (−5.77)	0.0001 (0.51)	−0.0002 (−1.24)	−0.0001 (−1.84)	
Sex_head						
Female	0.083*** (3.75)	−0.06*** (−5.47)	−0.115** (−2.01)	−0.228 (−1.84)	−0.057 (−1.87)	
Edu_head						
Through EGS/NFEC/AEC	0.0495 (0.38)	0.0737 (0.95)	0.1450 (0.62)	−0.075 (−0.17)	−0.0282 (−0.19)	
Through TLC	0.4214 (1.78)	0.051 (0.34)	0.266 (0.59)	−0.71 (−0.97)	0.118 (0.36)	
Others	0.116 (0.86)	0.0238 (0.31)	0.183 (0.78)	−0.01 (−0.02)	0.032 (0.22)	
Below primary	0.040 (1.53)	0.0045 (0.30)	0.0248 (0.44)	0.083 (0.70)	−0.176*** (−5.62)	
Primary	0.086*** (3.37)	0.0195 (1.37)	0.0764 (1.37)	0.0908 (0.73)	−0.088*** (−2.88)	
Middle	0.223*** (8.86)	0.046*** (3.36)	0.0686 (1.17)	0.0834 (0.65)	0.0086 (0.28)	
Secondary	0.452*** (16.80)	0.117*** (8.25)	0.0814 (1.22)	0.0149 (0.10)	0.0545 (1.59)	

(continued)

Table 6.1 (continued)

	LPG		Coke and coal		Dung cake	
Variable name	Probit	Truncated regression	Probit	Truncated regression	Probit	Truncated regression
Higher secondary	0.481*** (14.78)	0.15*** (8.97)	0.1421 (1.72)	0.1837 (0.98)	0.141*** (3.26)	
Diploma/ certificate course	0.739*** (11.22)	0.1451*** (5.17)	0	−0.532 (−1.10)	−0.35** (−2.33)	
Graduate	0.6314*** (17.24)	0.184*** (10.06)	0.2766*** (3.04)	0.2077 (1.06)	0.072 (1.36)	
Postgraduate and above	0.7206*** (13.41)	0.1584*** (6.60)	0.0036 (0.02)	0.4890 (1.18)	−0.0754 (−0.82)	
Social Group						
ST	−0.18*** (−7.60)	0.1667*** (12.73)	−0.1224** (−2.21)	0.2992** (2.29)	−1.25*** (−21.87)	
SC	−0.27*** (−12.23)	−0.0206 (−1.70)	−0.181*** (−3.65)	−0.0657 (−0.63)	0.128*** (5.02)	
OBC	0.008 (0.51)	0.0129 (1.66)	−0.128*** (−3.39)	0.1342 (1.63)	0.069*** (3.27)	
Religion						
Islam	−0.063*** (−2.95)	−0.086*** (−8.05)	0.0397 (0.86)	−0.0047 (−0.05)	−0.252*** (−9.02)	
Christianity	0.156*** (5.63)	0.0434*** (3.06)	−0.536*** (−4.63)	−0.1334 (−0.46)	−0.901*** (−6.74)	
Others	0.252*** (8.71)	0.1*** (7.30)	−0.281** (−2.61)	−0.542* (−1.99)	0.671*** (18.94)	
No. of observations	59,664	20,189	59,143	931	59,664	
LR chi^2	16691.79	3748.84	282.21	47.64	3843.34	
Pseudo R^2	0.2518	−13570.78	0.0391	−1314.87	0.1264	

Note: Figures in parentheses are t-value. '***', '**' and '*' denote 1%, 5% and 10% levels of significance, respectively.
Source: Author

the probit model reflect that both income and land assets have a direct effect on the selection of both LPG, and coke and coal, but only income has a negative effect on the selection of dung cake. This indicates that the advancement of the economic position of households reduces the probability of selection of dung cake for cooking fuels. However, income is a significant determinant of actual consumption of LPG in the truncated regression model. The results suggest that if LPG, coke and coal are chosen, then the amount of consumption of these fuels increases at a diminishing rate with income (as its square term is negative). Land holding (LO) positively affects the amount of LPG, and coke and coal consumption.

Household size has a positive impact on the selection of LPG and dung cake. However, it has a negative impact on the selection of coke and coal but has a positive impact on the consumption of LPG. This implies that members of households raise the demand for cooking fuels. For the average age of the household, we observe a positive impact on the selection of coke and coal as cooking fuel. It is observed that the probability to choose coke and coal increases at a falling rate with an increase in the average age of households, whereas the probability to choose LPG and dung cake decreases at an increasing rate with an increase in the average age of households. The impact of the age of household head is similar for the selection of firewood, but the results are opposite for the selection of LPG and dung cake. The age of household head also significantly determines the amount of LPG consumption.

The education level of households also plays an important role in shaping the cooking fuel choice decision of households. First, the average educational level of households has a negative impact on the selection of dung cake (comparatively dirty fuels), but has a positive impact on the selection of LPG (comparatively clean fuel). It is also true in truncated regression. Hence, education plays a major role in the choice decision as well as consumption quantity decision of cooking fuels. The household head is the major decision-maker of the household, so their educational qualification might play an important role in the fuel consumption decision of the household. Results suggest that at lower levels of educational attainment of household head, the probability to choose dirty fuel (dung cake) is high and to choose clean fuel (LPG) is low, whereas at the higher educational attainment of household head, the probability to choose LPG is high. The results of the quantity hurdle also reveal a similar picture. For dung cake, the amount of consumption equation is not estimated due to the non-availability of its quantity. For LPG, the amount of consumption increases throughout all educational levels of household head. This implies that the educational level of household head as well as other members encourages clean fuel consumption while discouraging the consumption of dirty ones.

The results suggest that for LPG, the probability of selection is greater in the case of female-headed households compared to that of male, whereas for coke and coal, the probability to choose is less for female-headed households compared to that of male. The results of the truncated regression model reflect a decline in consumption of LPG, coke and coal for female-headed households compared to male-headed households.

The results reveal that the three major social groups, namely ST, SC and OBC, have a positive impact on the selection of firewood as the cooking fuel compared to the general caste households. The estimates of the second hurdle state that if firewood is chosen, then only ST households make a positive amount of consumption, whereas the other two social groups report lesser amount of consumption compared to general caste households. The results reflect that the probability of selection of LPG, coke and coal is low for ST and SC households and low in the case of dung cake only for ST households, compared to the general caste households. If LPG and coal are chosen, then ST households report a positive amount of consumption compared to general caste households.

6.3 Section B: Firewood and Chips

The previous chapter suggested that the most preferred fuel is firewood and chips in rural India. Now, this section exclusively focuses on all possible determinants of the consumption of firewood and chips in rural India. Out of the huge set of biomass fuels, this study restricts our attention only to firewood and chips.[1] Here, we conduct an empirical investigation in detail and identify the role of different social and economic factors that influence the consumption of firewood and chips in rural India. Poor economic condition is associated with consumption of dirty fuels like firewood and chips. It is more likely that households shift their consumption from dirty to cleaner fuels (say, firewood and chips to LPG) due to increment in income.[2] This fuel switching might affect due to other factors such as education, family size, land holdings, electrification and occupational patterns in rural India.

It has been argued that households may avail firewood and chips as a free item. It may also be consumed from home-grown products, which may be a by-product of agricultural production. With the limited information, we now explore the role of factors mentioned earlier in determining the choice of firewood and chips. Applying probit and truncated regression models, we estimate the marginal effects of socio-economic variables on the consumption of firewood and chips. Table 6.2 provides the estimated marginal effects of social and economic variables on firewood and chips consumption. The major socio-economic factors are statistically significant at 5% level and/or above, except income level (i.e. LogMPCE) in all three models. The coefficient of income square (LogMPCE^2) is negative and that of land is positive and both are highly significant in all possible models. The results of Table 6.2 suggest that significant determining factors are land, family size and its composition, level of education, fuel prices, occupations, social caste and religion, regional groups etc.

The data sets contain information on the amount of consumption incurred for firewood and chips and also on the sources related to this consumption. The source of consumption broadly comes from home-grown stocks or/and free collection from the common property recourses or purchase. We shall explore the role of different socio-economic factors that determin the sources of consumption of this item. For example, it is likely that households with higher income or having more lands either decide not to consume this material, or if consumed depend on their home-grown

[1] It is the principal source of cooking fuel of the majority (>65%) of rural Indian households.

[2] Income plays an important role for climbing above the energy ladder model (Heltberg, 2004). The first phase is characterised by reliance on biomass; in the second phase, households move to transition fuels like kerosene and the final phase is to switch to cleaner fuels like LPG and electricity for cooking purposes.

Table 6.2 *Marginal effects* of social and economic variables on total firewood consumption

	Probit model			Truncated regression model		
Variables	M1	M2	M3	M1	M2	M3
Log(MPCE)	0.12 (0.06)	0.23 (0.06)	0.24 (0.06)	1.41 (0.12)	1.24 (0.11)	1.27 (0.11)
Log(MPCE)^2	−0.03** (0.01)	−0.03** (0.01)	−0.03** (0.01)	−0.13** (0.01)	−0.11** (0.01)	−0.12** (0.01)
Log(land owned)	0.02*** (0.00)	0.02*** (0.00)	0.02*** (0.00)	0.01*** (0.00)	0.01*** (0.00)	0.02*** (0.00)
Family size		0.02*** (0.00)	0.02*** (0.00)		−0.09*** (0.00)	−0.09*** (0.00)
Age of the head of family		0.00*** (0.00)	0.00*** (0.00)		0.00*** (0.00)	0.00*** (0.00)
% of child		−0.08** (0.02)	−0.07** (0.02)		−0.16** (0.04)	−0.15** (0.04)
% of adult female		0.08** (0.02)	0.07** (0.02)		0.10** (0.04)	0.08** (0.04)
% of adult male		0.04** (0.02)	0.04** (0.02)		0.10** (0.04)	0.09** (0.04)
Log(100*price of firewood)	−0.11** (0.01)	−0.10** (0.01)	−0.10** (0.01)	−0.71** (0.01)	−0.67** (0.01)	−0.67** (0.01)
Log(100*price of kerosene)	−0.08** (0.01)	−0.06** (0.01)	−0.06** (0.01)	0.08** (0.03)	0.03** (0.03)	0.04** (0.03)
Log(100*price of LPG)	−0.04** (0.03)	−0.03** (0.03)	−0.03** (0.03)	−0.10** (0.05)	−0.12** (0.05)	−0.12** (0.05)
Education (base: below primary)						
Primary		−0.00** (0.01)	−0.00** (0.01)		−0.04** (0.02)	−0.04** (0.02)
Secondary/diploma		−0.04** (0.01)	−0.03** (0.01)		−0.10** (0.02)	−0.10** (0.02)
Graduate and above		−0.15** (0.01)	−0.14** (0.01)		−0.32** (0.03)	−0.31** (0.03)
Occupation (base: non-agriculture)						
Agri labour			0.03** (0.01)			0.07** (0.02)
Self-employed			0.04** (0.01)			−0.01** (0.01)
Social group (base: general caste)						
ST		0.13** (0.01)	0.13** (0.01)		0.27** (0.02)	0.27** (0.02)
SC		0.10** (0.01)	0.10** (0.01)		0.07** (0.02)	0.07** (0.02)
OBC		0.06** (0.01)	0.07** (0.01)		0.03** (0.02)	0.03** (0.02)
Perform ceremony (base: yes)						

(continued)

Table 6.2 (continued)

	Probit model			Truncated regression model		
Variables	M1	M2	M3	M1	M2	M3
No		−0.04** (0.02)	−0.04** (0.02)		−0.18 (0.06)	−0.18 (0.06)
Religion (base: other group)						
Hindu		0.07** (0.02)	0.07** (0.02)		0.22** (0.04)	0.22** (0.04)
Muslim		0.06** (0.02)	0.07** (0.02)		0.26** (0.04)	0.27** (0.04)
Christian		0.10** (0.02)	0.11** (0.02)		0.21 (0.05)	0.21 (0.05)
Regional group (base: Western)						
BIMAROU		−0.01** (0.01)	−0.00** (0.01)		0.18** (0.02)	0.20** (0.02)
Northeastern		−0.05** (0.01)	−0.04** (0.01)		0.15** (0.02)	0.16** (0.02)
Northern		0.02** (0.01)	0.03** (0.01)		−0.11** (0.02)	−0.11** (0.02)
Southern		−0.00** (0.01)	−0.00** (0.01)		0.03** (0.02)	0.04** (0.02)
Sex of household head (base: male)						
Female		0.03** (0.01)	0.03** (0.01)		0.09** (0.02)	0.10** (0.02)
Principle source of lighting (base: others)						
Electricity		−0.00** (0.01)	−0.00** (0.01)		−0.01** (0.01)	−0.01** (0.01)
Diagnostics						
Wald test	1408.15	2355.07	2366.02	4241.05	7693.68	7791.94
Pseudo R^2	0.11	0.15	0.15	0.10	0.17	0.17

Note: Figures in parentheses are p-values. '***', '**' and '*' denote 1%, 5% and 10% level of significance, respectively.
Source: Author

products. In order to address this question, a multinomial logit model[3] is used for estimation purposes. The marginal logit model tells only one part of the story. It is

[3] Applications of discrete choice model like multinomial logit/probit model in this literature is very common. For example, Jumbe and Angelsen (2011) in the context of Malawi had explored the role of different factors effecting the choice of sources of firewood collected from different types of forests, namely forest reserves, customary forests and plantation forests. Rao and Reddy (2007), in the context of India, explored the factors that influence individual's decision to choose a particular energy carrier, viz., firewood, LPG, kerosene and electricity. Heltberg (2004, 2005) studied the fuel-switching behaviour of households, ranging from only solid fuels to partially solid and non-solid fuels and finally non-solid fuels. There are plenty of examples; for further references, go through the papers cited in these papers.

not clear whether these factors determine the choice of fuel source and also explain the intensity of consumption of this material in the same direction.

For example, considering the factor 'land', in the multinomial probit model, we have observed that its marginal effect is positive when the source is home-grown production. The same effects, on the other hand, turn to be negative when the source is free collection or purchase. It is likely that households having land can fulfil most of their demands from home-grown products and is not necessary to purchase or even collect as a free commodity. However, those households having a larger amount of land may not fulfil their demand for firewood and chips solely on the basis of home-grown production, since firewood and chips often is necessary for different agricultural purposes.[4] Thus, the amount of firewood purchased or obtained freely is expected to be positively related to land owned.

A probit model explains the probability of different factors on whether the household makes a positive consumption or not. A truncated regression model explains the role of the same factors for the intensity of consumption of the said item. We estimate the effects of different factors on the total amount of consumption, from home grown only, from purchase and lastly on amount of free collections (Table 6.3). Empirical evidence from the above-said two models shows that the coefficient of land indeed changes sign (when firewood is solely based on purchase or free collection) and supports our expectations as stated in the earlier section.

6.4 Major Findings

We report the marginal effects of different socio-economic variables on source-wise firewood and chips consumption in rural India in Table 6.3. The coefficient of land is again significant but the sign of its coefficients is mixed. Major results are discussed below.

Land owned: Land plays an important role in explaining many hidden facts of the data. Recall that in Fig. 5.2, we have observed that total consumption increases with increment in land. Our empirical findings (following Table 6.3) show that this factor explains not only the probability of firewood consumption but also its intensity in a positive direction in the case of home-grown and free collections. On the contrary, when we shift our attention to the firewood consumption only from purchase, the same factor works in a negative direction, which is justified as land size increases, the purchase of firewood declines. It implies that those who have more land probably fulfil their requirements from the home-grown sources. However, the probit model provides a significant result, and the truncated model ends with a significant coefficient for the variable land. Firewood consumption from home-grown sources increases with the increment in land. In

[4] Firewood is used not only for cooking but also for the purpose of water heating, space heating, lighting, etc. (Andreae, 1991).

Table 6.3 Source-wise (home grown, purchase and free collections) *marginal effect* of factors of firewood and chips

	Probit			Truncated regression		
Variables	Home grown	Purchase	Free collections	Home Grown	Purchase	Free Collections
LogMPCE	0.28 (0.06)	0.49 (0.09)	0.21 (0.09)	1.32 (0.19)	1.11 (0.20)	1.77 (0.26)
$LogMPCE^2$	−0.02*** (0.01)	−0.05*** (0.01)	−0.04*** (0.01)	−0.12** (0.02)	−0.09** (0.02)	−0.18** (0.03)
Logland owned	0.05*** (0.00)	−0.02*** (0.00)	−0.00*** (0.00)	0.06*** (0.01)	−0.00*** (0.01)	0.02*** (0.00)
Family size	0.02*** (0.00)	0.00*** (0.00)	−0.01*** (0.00)	−0.07*** (0.01)	−0.09*** (0.01)	−0.11*** (0.01)
Age of head of family	0.00*** (0.00)	0.00*** (0.00)	−0.00*** (0.00)	−0.00** (0.00)	0.00** (0.00)	0.00** (0.00)
% of child	−0.02** (0.02)	0.03** (0.02)	−0.06** (0.02)	−0.21 (0.07)	−0.24 (0.06)	−0.04 (0.06)
% of adult female	−0.01** (0.02)	0.06** (0.02)	0.03** (0.02)	0.05 (0.08)	0.10 (0.06)	0.08 (0.06)
% of adult male	−0.03** (0.02)	0.05** (0.02)	0.02** (0.02)	0.09 (0.08)	0.09 (0.07)	0.04 (0.06)
Log(100*price of firewood)	−0.06*** (0.01)	0.13*** (0.01)	−0.15*** (0.01)	−0.60** (0.02)	−0.72** (0.02)	−0.65** (0.02)
Log(100*price of kerosene)	−0.11*** (0.01)	0.04*** (0.01)	−0.02** (0.02)	0.09** (0.05)	−0.02** (0.04)	0.10** (0.05)
Log(100*price of LPG)	−0.13** (0.03)	0.14** (0.03)	−0.05** (0.03)	−0.26 (0.10)	0.18 (0.08)	−0.45 (0.11)
Education (base: below primary)						
Primary	0.03*** (0.01)	0.02*** (0.01)	−0.04*** (0.01)	−0.05** (0.04)	−0.05** (0.03)	−0.07** (0.03)
Secondary/diploma	0.04*** (0.01)	0.03*** (0.01)	−0.09*** (0.01)	−0.17** (0.03)	−0.13** (0.03)	−0.06** (0.02)
Graduate and above	0.02*** (0.01)	0.01*** (0.01)	−0.15*** (0.01)	−0.38** (0.05)	−0.34** (0.05)	−0.19** (0.04)
Occupation (base: non-agriculture)						
Agri labour	0.00*** (0.01)	−0.03*** (0.01)	0.05*** (0.01)	0.17** (0.04)	0.05** (0.03)	0.08** (0.02)
Self-employed	0.06*** (0.01)	−0.07*** (0.01)	0.03*** (0.01)	0.06** (0.02)	−0.03** (0.03)	0.01** (0.02)
Social group (base: general caste)						
ST	−0.05*** (0.01)	−0.07*** (0.01)	0.23*** (0.01)	0.30** (0.04)	0.20** (0.04)	0.27** (0.03)
SC	0.00*** (0.01)	0.01*** (0.01)	0.11*** (0.01)	0.12** (0.04)	0.03** (0.03)	0.06** (0.03)
OBC	0.01*** (0.01)	0.01*** (0.01)	0.05*** (0.01)	0.06** (0.03)	0.02** (0.03)	0.02** (0.03)

(continued)

Table 6.3 (continued)

	Probit			Truncated regression		
Variables	Home grown	Purchase	Free collections	Home Grown	Purchase	Free Collections
Perform ceremony (base: yes)						
No	0.04** (0.02)	−0.04** (0.03)	−0.07 ** (0.03)	−0.23 (0.08)	−0.18 (0.08)	−0.14 (0.12)
Religion (base: other group)						
Hindu	0.01** (0.02)	−0.01** (0.02)	0.11** (0.02)	0.12 (0.08)	0.12 (0.06)	0.35 (0.06)
Muslim	0.02** (0.02)	0.02** (0.02)	0.07 ** (0.02)	0.13 (0.08)	0.14 (0.07)	0.34 (0.07)
Christian	0.07** (0.02)	−0.01** (0.03)	0.12** (0.02)	0.17 (0.09)	0.11 (0.08)	0.25 (0.08)
Regional group (base: Western)						
BIMAROU	0.08** (0.02)	0.21** (0.02)	−0.27*** (0.01)	0.17** (0.04)	0.16** (0.05)	0.23 ** (0.04)
Northeastern	−0.08*** (0.01)	0.00*** (0.01)	0.02 *** (0.01)	0.12** (0.05)	0.07** (0.05)	0.28 ** (0.03)
Northern	−0.08*** (0.01)	0.11*** (0.01)	0.01 *** (0.01)	−0.01** (0.04)	−0.07** (0.04)	−0.19 ** (0.03)
Southern	0.01*** (0.01)	0.10 *** (0.01)	−0.10 *** (0.01)	0.00** (0.04)	0.08** (0.04)	0.00 ** (0.03)
Sex of the household head (base: male)						
Female	0.01*** (0.01)	0.01*** (0.01)	0.00*** (0.01)	0.05** (0.04)	0.13** (0.03)	0.04** (0.03)
Principle source of lighting (base: others)						
Electricity	−0.04*** (0.01)	0.00*** (0.01)	0.02*** (0.01)	0.03** (0.03)	−0.02** (0.03)	−0.03** (0.02)
Diagnostics						
Wald test	2698.81	1104.74	2259.09	2029.74	2518.62	3322.54
Pseudo R^2	0.15	0.06	0.12	0.13	0.17	0.18

Note: Figures in parentheses are p-values. '***', '**' and '*' denote 1%, 5% and 10% level of significance, respectively.
Source: Author

the case of free collection, this factor works in a negative direction, while this result is completely inverted once we move to the intensity of consumption.

Prices: The price of firewood is negatively related to firewood and chips consumption in the case of home-grown and free collection in both probabilistic and intensity terms. The price of firewood is positive in probit and negative in intensity terms. The effects of prices of the other two items, namely kerosene and LPG, is not straightforward. In the context of probit model, kerosene is significant and negative. In the truncated regression model, the same factor is insignificant, but the price of LPG plays a negative role in the consumption of firewoods. On the contrary, when we focus only on the purchase of this item, the table shows that firewood, kerosene and LPG price play a positive role in consumption. It implies that as a result of the increment in the price of any of the above-mentioned commodity, the probability that a household will purchase this item increases. The intensity of purchase, however, is negatively related to its purchase. Although kerosene prices turn out to be insignificant, it is clearly observable that as a result of increment in LPG prices, firewood consumption increases. For the other two sources, all prices in general are negatively related to firewood consumption. There are some exceptions: for example, as a result of increment in kerosene prices, the marginal effect of increment in firewood is 0.1.

Education: In general, a higher level of education implies a decline in firewood consumption, for almost all the sources. In households having at least one member above primary education level, the probability of firewood consumption reduces in the case of free collection, while it rises in the case of home grown and purchase. Education above primary levels is significantly reducing the incidence of firewood consumption in home grown, purchase and free collections.

Occupations: Occupations of labour of household are directly related to firewood and chips consumption in the case of home-grown and free collection; however, an inverse relation is observed for self-employed in the case of its purchase. The probability and incident of firewood consumption decline in the case of purchase for self-employed compared to non-agriculture labourer.

Social groups: All the social groups have a direct relation with all sources of firewood and chips consumption incident; however, only scheduled tribes select less firewood and chips.

Perform ceremony: Rural households in India select all sources of firewood and chips for their ceremony performance.

Sex of household heads: It is directly related to all three sources of firewood and chips consumptions. The probability and incident of firewood consumption of female-headed households increase in comparison with male-headed households in rural India.

Principle source of lighting: It is inversely related to firewood consumption if home grown otherwise direct relations. The probability of home-grown firewood consumption declines due to the use of electricity for lighting.

Regional group: This has both positive and negative relations with sources of firewood consumptions. The BIMAROU, Northeastern and Southern regions show direct relation to firewood consumption except North-East in the case of home grown consumption. However, Northern region is inversely related to it.

6.5 Conclusion

This chapter has discussed the estimated results of the selection of fuels and estimated the determinants of the amount of consumption using probit and truncated regression models. The probit model estimates the determining factors for the probability of selection of fuels for cooking purposes, whereas the truncated regression model estimates the determinants of consumption amount. However, there exists a distinction between desired and actual consumption. There might be some households who report a positive desire for the consumption of cooking fuel, but their actual consumption of that cooking fuel is zero. There might be certain factors that inhibit the carrying out of the change for actual consumption to be positive. This is also reflected by the number of observations in both the models. The number of observations is quite large in the case of probit model for fuel selection compared to that in the case of truncated regression model. A huge number of observations are missing in the estimation process. To overcome such limitation, the double hurdle model is appropriate and applied in the following study to estimate the determinants of firewood and chips (most widely used dirty for cooking fuel) consumption in rural India. The estimation of double hurdle model does not only reflect any loss in the observations but also incorporate the socio-economic characteristics of those households which were earlier excluded by the truncated regression model for reporting zero consumption. The double hurdle model might provide better results than the probit and truncated regression models.

Chapter 7
Determinants of Firewood and Chips Consumption in Rural India: Double Hurdles Model Approach

7.1 Introduction

Households take decisions regarding the selection of cooking fuel and its quantity consumption. Here, household decisions are in two parts: the first part of the decision is to select cooking fuel among available alternative fuels, and the second part of the decision is to determine the level of the selected cooking fuel. So, household decision-making factors associated with cooking fuel choice and quantity consumption are crucial research issues. The household's decision-making factors may vary from the first stage to the second stage. The determining factors of cooking fuel choice and consumption might be important indicators which focus on certain targets of Sustainable Development Goal 7 (SDG7). The determinants of cooking fuel choice at the household level are important research issues focusing on Sustainable Development Goals (SDGs) for the year 2030. Researchers have not highlighted the issue of hurdles of cooking fuel choice in rural parts of developing economies like India. Earlier studies mainly focused on and discussed prices of fuels and socio-economic variables. However, no study highlights non-price issues associated with cooking fuel choice and consumption, except Daripa and Dinda (2024). This chapter, following Daripa and Dinda (2024), attempts to identify the non-price factors of cooking fuel choice of household decision.

Rural households use traditional *dirty* fuels (such as firewood and chips, dung cakes, crop residues and coal) which generate emissions in the process of their combustion. Modern *clean* fuels (such as biogas, LPG and electricity) are more efficient in terms of generation of pollution than traditional fuels. CO_2 emissions are added

---

This chapter is based on (a) Daripa and Dinda (2022): Hurdles in fuel choice and consumption in rural India. In Bagli et al. (eds.), *Persistent and Emerging Challenges to Development*, India Studies in Business and Economics, https://doi.org/10.1007/978-981-16-4181-7_6 and (b) Daripa and Dinda (2024): Stage-based non-price determinants of fuel choice and its consumption decision of households in rural India. *The Indian Economic Journal* (*forthcoming*). https://doi.org/10.1177/00194662241238589

S. Daripa, S. Dinda, *Hurdles in Fuel Choice in Rural India*, SpringerBriefs in Geography,
https://doi.org/10.1007/978-3-031-71605-8_7

to the atmosphere due to the combustion of firewood and chips. A large part of the supply of firewood and chips is associated with forest degradation or deforestation that threatens the environment as well as civilised human society. Emerging countries have to face a world of uncertainty that leads to unsustainable development.

In this context, it is an important requirement for understanding and identifying the determinants of cooking fuel consumption decision at the household level. Now, attention is needed on hurdles or difficulties faced by households in their fuel choice and quantity consumption decision in emerging economies such as India. This study mainly focuses on fuel for cooking purposes in rural areas of India. Literature mostly reports and discusses fuel price, not on non-price factors of social issues. Research study is required to highlight and focus on non-price factors such as land, composition of household like family size, family age composition, average family education level; social groups in terms of caste and religion; and the characteristics of decision-maker of households such as family head's age, sex, level of education, etc. Land is a crucial asset of rural households in India. A wealth of a household is measured in terms of land assets, which also indicate the socio-economic status of rural households in India (Daripa & Dinda, 2022, 2024). Here, a question arises about how socio-economic backgrounds of households influence their fuel choice and consumption decision. What are the factors responsible for making fuel choice decision of households in rural India? Following Daripa and Dinda (2022, 2024), this chapter examines and identifies stage-based determinants of non-price factors of cooking fuel choice and consumption decision of households in rural India.

From preliminary data analysis, we have observed that only 37,410 households participated in the selection of firewood and chips from ten alternative available cooking fuels in rural India. However, actually, 32,337 rural households have used a positive amount of firewood and chips for their cooking purposes. So, 5073 rural households have responded to participate in firewood and chips selection at the first stage, but they actually consumed 'zero' amount of firewood and chips at the second stage. Here, we realised a difference between their desired and actual consumption of firewood and chips. Definitely, certain factors are responsible for such behaviour of the above said households. In this context, the double hurdle model of Cragg (1971) is applicable and suitable to investigate hurdles in the selection of fuel and its quantity consumption at the household level.

Various studies[1] have pointed out several determining factors of fuel consumption at the household level like income, education, age, caste, religion, sex, educational level of household head, land, etc. In this context, we have considered land as the determining factor of fuel choice and also analysed it without land. The presence of land may affect other variables. So, in this chapter, the estimated results are discussed in two sub-sections focusing on (i) results with land and (ii) without land.

The double hurdle model (Cragg, 1971) estimates two different models in two stages: (i) the probit model is used in the first stage and (ii) the tobit model is used

[1] See Gupta and Kohlin (2006), Gundimeda and Kohlin (2008), Couture et al. (2012), Jingchao and Kotani (2012), Daripa and Dinda (2022), etc.

in the second stage. In the first stage, it estimates the probability of fuel selection out of available ten alternative fuels. This is the participation equation, which identifies the determining factors of choice of firewood and chips at the household level. After fuel item selection, a question arises about the quantity consumption. So, in the second stage, determining factors of actual firewood and chips consumption is estimated in the quantity equation.

Table 7.1 provides the estimated results of participation and quantity equations. The results of column A and column B in Table 7.1 display the results of the double hurdle model (DHM) for firewood and chips selection and consumption decision in rural India, considering with and without land assets. The only difference between both columns A and B (in Table 7.1) is the land asset. Land ownership (LO) is included in column A in Table 7.1, while LO is excluded in column B in Table 7.1.

Table 7.1 Estimated double hurdle model with and without land

	A (with land assets)		B (without land assets)	
Variables	Participation equation	Quantity equation	Participation equation	Quantity equation
Constant	0.6485 (0.63)	1.5184 (1.94)	**3.4965***** **(5.26)**	**1.7458***** **(3.69)**
LogPMCE	0.1011 (0.37)	**0.6492***** **(3.08)**	**−0.6006***** **(−3.47)**	**0.5676***** **(4.51)**
LogMPCE_sq	−0.0296 (−1.65)	**−0.0329**** **(−2.32)**	0.0207541 (1.84)	**−0.02839***** **(−3.38)**
LO	***36.2663****** ***(12.09)***	***-23.158****** ***(-9.32)***	–	–
Hhsize	**0.0209***** **(3.03)**	**0.0614***** **(12.85)**	**0.0419***** **(10.42)**	**0.0525***** **(20.95)**
Age	**0.0128***** **(3.22)**	**0.0087***** **(3.07)**	**0.0149***** **(5.3)**	**0.0067***** **(3.53)**
Age_sq	**−0.00014***** **(−2.97)**	**−0.0002***** **(−6.19)**	**−0.0002***** **(−4.76)**	**−0.0002***** **(−7.85)**
Edu	**−0.061***** **(−9.04)**	**0.0196***** **(3.9)**	**−0.0682***** **(−14.79)**	**0.0136***** **(4.24)**
Age_head	**0.0196***** **(4.4)**	−0.0012 (−0.35)	**0.0212***** **(6.91)**	−0.0011 (−0.5)
Age_head_sq	**−0.0002***** **(−4.57)**	0.00005 (1.64)	**−0.0002***** **(−6.99)**	0.00004 (1.91)
Sex_head				
Female	**0.1384***** **(4.95)**	**−0.057***** **(−2.9)**	0.0156 (0.75)	**−0.0413***** **(−2.92)**
Edu_head				
Through EGS/NFEC/AEC	−0.0794 (−0.48)	0.0079 (0.07)	−0.15296 (−1.45)	0.0507 (0.7)
Through TLC	−0.1463 (−0.42)	−0.2044 (−0.82)	0.1516 (0.61)	−0.2668 (−1.8)

(continued)

Table 7.1 (continued)

	A (with land assets)		B (without land assets)	
Variables	Participation equation	Quantity equation	Participation equation	Quantity equation
Others	−0.16688 (−1.07)	0.043 (0.37)	−0.059 (−0.51)	0.0279 (0.37)
Below primary	**0.141***** **(4.15)**	0.031 (1.37)	**0.1389***** **(5.72)**	**0.0465***** **(3.12)**
Primary	**0.087**** **(2.58)**	**0.0474**** **(2.03)**	**0.1224***** **(5.18)**	**0.0444***** **(2.94)**
Middle	**0.096***** **(2.75)**	0.0071 (0.29)	**0.1158***** **(4.87)**	−0.022 (−1.43)
Secondary	**−0.2175***** **(−5.66)**	−0.055 (−1.89)	**−0.1078***** **(−4.24)**	**−0.0613***** **(−3.47)**
Higher secondary	**−0.3093***** **(−6.43)**	0.064 (1.68)	**−0.1945***** **(−6.3)**	**−0.0471**** **(−2.11)**
Diploma/certificate course	**−0.5478***** **(−5.85)**	−0.0315 (−0.35)	**−0.3472***** **(−5.62)**	**−0.1639***** **(−3.11)**
Graduate	**−0.586***** **(−10.58)**	**0.1988***** **(4.21)**	**−0.366***** **(−10.55)**	−0.0097 (−0.37)
Postgraduate and above	**−0.6762***** **(−8.42)**	**0.1818**** **(2.39)**	**−0.4694***** **(−9.37)**	−0.0689 (−1.63)
Social group				
ST	**0.352***** **(9.72)**	**0.12***** **(4.76)**	**0.4212***** **(18.17)**	**0.1553***** **(10.61)**
SC	**0.2879***** **(10.54)**	**−0.1942***** **(−9.51)**	**0.1588***** **(8.11)**	**−0.1252***** **(−9.23)**
OBC	**0.251***** **(11.1)**	**−0.2332***** **(−13.23)**	**0.1705***** **(11.47)**	**−0.1858***** **(−17.31)**
Religion				
Islam	−0.0363 (−1.33)	**0.0766***** **(3.83)**	**−0.0874***** **(−4.53)**	**0.0947***** **(7.06)**
Christianity	0.0357 (0.84)	**0.1554***** **(5.08)**	**0.0129** **(0.45)**	**0.2004***** **(11.04)**
Others	**−0.4864***** **(−11.74)**	**0.1889***** **(5.21)**	**−0.355***** **(−12.79)**	**0.1617***** **(7.42)**

Note: '***' and '**' denote 1%, and 5% level of significance, respectively.
Source: Daripa and Dinda (2022, 2024)

7.2 Determinants of Firewood Consumption of Household

The results of the double hurdle model are reported in Table 7.1. The statistically significant coefficients of variables in the participation equation (A part of Table 7.1) are land owned (LO), household size, age, age-square, education level of household, age of household head and its square, female-headed household, social caste (ST, SC and OBC), 'others' religion group and educational level of household head at below primary level, above middle level and primary level of education of household head. The coefficients of income and its square are statistically insignificant in the participation equation in the presence of land (LO); however, the coefficient of land is highly significant (A part of Table 7.1). It suggests that land is a deterring factor, but income has no role in the selection of firewood and chips in the presence of land[2] in rural India. Land ownership may provide insightful probable information regarding choice of fuel consumption, particularly firewood and chips, at the household level in rural area. The positive coefficient of land holding (LO) in participation equation indicates that an increase in land holding of households raises its probability to select firewood and chips as cooking fuel in rural India. This result possibly suggests that LO leads to increased production of firewood and chips, thereby the probability of its consumption choice increases. Negative coefficient of LO in quantity equation indicates that the quantity of firewood and chips consumption reduces with land holding, which indicates the possibility to switch over to alternative fuels that are accessible or/and available in rural India.

The estimated results of quantity equation (a part of Table 7.1) shows that log monthly per capita consumption expenditure (MPCE), land owned, average educational level of household, average age of household and its square, female-headed household, household size, all social and religious groups and graduate household head are statistically significant at 1% level while the square of log MPCE, education level of household head at the primary level and post graduate level are significant at 5% level.

Coefficients of MPCE in A part of Table 7.1 are statistically highly significant in quantity equation but insignificant in participation equation. The result of participation equation suggests that income is probably not a determining factor in the selection of firewood and chips consumption in rural India. However, a statistically significant positive sign of the coefficient of log MPCE and negative sign of its square in quantity equation suggest that the amount of firewood consumption increases with income at a declining rate. These results suggest that income has no significant role in firewood selection; however, if it is selected, then income crucially determines its consumption quantity.

The coefficients of household size in participation and quantity equations are positive and significant. The probability of firewood and chips selection and its consumption quantity rises with family size. It indicates that increased manpower probably helps to collect firewood and chips nearly free of cost in rural India.

The coefficient of 'age' and its square in the participation equation are, respectively, positive and negative that indicate the probability of the selection of firewood and chips rises with an average age of the household at a declining rate. The results of quantity equation suggest that firewood and chips consumption amount increase with the average age of households at a decreasing rate, if the household chooses to consume it in the first stage. Similarly, the coefficients of the age of household head and its square are, respectively, positive and negative in the participation equation

[2] The results of participation equations in Table 7.1 are distinctly clear and indicate possible association between income and land in rural India.

that suggests probability of firewood choice increases with the falling rate with the age of household head.

The probability of choice of firewood and chips reduces with the education level of household, but they consume positive quantity if it is selected. In this context, the educational level of the household head is crucial for the choice of firewood and chips and its actual consumption. Negative coefficients of secondary and higher level of education of household head in participation equation indicate that the probability of firewood and chips selection declines with the level of education of household head. Positive coefficients of the educational level of the household head at below primary, primary and middle school level indicate that the probability of firewood and chips selection increases at a lower level of education of household head. Quantity equation reflects slightly different pictures. Primary educated household heads increase the consumption of firewood and chips; however, the same results also hold in the case of graduate and above level of education of household head. It is probably due to no accessibility or supply constraint of clear fuels. Surprisingly, the result of household head's secondary level of education indicates that the household head desires to reduce firewood and chips consumption. These findings suggest that policymakers should focus on the educational level of household heads beyond the secondary level as well as the need to ensure the availability of alternative clean fuels matching with desired demand.

The coefficients of female-headed households are positive in the participation stage and negative in the quantity selection stage. So, the probability of selecting firewood and chips rises with female-headed households in the participation equation while the amount of firewood and chips consumption reduces in the second stage compared to male-headed households. These results possibly indicate that female-headed households might be financially worse off and select cheap firewood and chips compared to male-headed households. Females are more aware of the ill effects of using firewood and chips as cooking is in women's domain in rural India (Daripa & Dinda, 2024). As per the quantity equation, this might be the possible reason for less firewood and chips consumption compared to male-headed households in rural areas.

The double hurdle model represents the effect of different social groups, namely ST, SC and OBC, compared to the general caste on the fuel selection behaviour of households. The coefficients of social groups (namely ST, SC and OBC) are positive in the participation equation. The probability of choice of firewood and chips rises for deprived social groups (ST, SC and OBC) compared to others. The coefficient of ST is positive and the coefficients of SC and OBC are negative in the quantity equation. It indicates that ST households consume more and more quantity of firewood and chips, while SC and OBC households reduce the amount of firewood and chips consumption.

In the participation equation, only others group reduces the probability of the selection of firewood and chips. Quantity equation results of religion group indicate that Islam, Christianity and others raise the consumption of firewood and chips.

7.3 Determinants of Firewood Consumption of Households Without Land Assets

Land assets are associated with household income in rural India. Land assets or land-owned (LO) variable is dropped from the above-said model due to probable multicollinearity issue in econometric exercise and re-estimate the double hurdle model for the analysis purpose. Part B of Table 7.1 presents the estimated results of double hurdle model excluding land assets (LO), while part of A of Table 7.1 reports results including LO. Part B of Table 7.1 provides the estimated results of participation and quantity equations for the selection of firewood and chips, and its quantity consumption decision without land ownership (LO). The results of participation equation in part B of Table 7.1 show that income is highly significant and an important determining factor of the selection and consumption of firewood and chips in participation and quantity equations in the absence of land. Part B of Table 7.1 provides that family composition variables like family size, age and education are significant determining factors of firewood and chips. Social castes (ST, SC and OBC) and religions are also significant determining factors in both decision stages. Female-headed household determines less amount of firewood and chips consumption in the quantity equation, if it is selected though the participation equation does not indicate it. However, the above and below primary education level of household heads significantly determines the selection and amount of consumption of firewood and chips. The absence of land signifies income as one of the determining factors, otherwise, land determines the choice of firewood and chips of rural households in India. Multicollinearity issue does not arise here. Again, we also find other social factors highlighting characteristics of family composition and sex and educational quality of household heads.

The results of the participation equation in part B of Table 7.1 suggest that the probability of selection of firewood and chips reduces with income. The quantity equation indicates that firewood and chips consumption amount rise with a declining rate in the second stage, if firewood and chips is selected in the first stage.

The results of the head's educational levels, social and religious groups, and other variables are mostly similar to part A of Table 7.1. The result of female-headed households is insignificant in the participation equation only, otherwise, it also supports earlier results.

Comparing the results of the double hurdle model for estimating the determinants of firewood consumption as the major cooking fuel in Table 7.1 (i.e. excluding and including land assets), it is observed that the variables used in the study in both cases resemble a similar picture except the variable referring to income of the household denoted by LogMPCE and female-headed household. The results reflect that in the participation equation of part B of Table 7.1 (without LO), income significantly affects the choice of firewood and chips as the major cooking fuel and has a negative impact on its selection, but it is observed in part A of Table 7.1 (with LO) that income does not significantly affect the choice of firewood and chips as the major cooking fuel. The quantity equation in both the cases reflect a similar picture stating that income leads to an increase in the amount of consumption of firewood and chips in both cases. The results also reflect that female-headed household determines the selection of firewood as cooking fuel in rural India in the case of with LO but not without LO (see Table 7.1). It can be inferred from both the cases that land holding is an important aspect reflecting the asset structure of rural households. Hence, inclusion of LO in the model gives us a clear picture than excluding LO from the model.

7.4 Conclusion

This chapter has raised and discussed the research issue of double hurdles in fuel choice and consumption decision of rural households in India. This study actually highlights stage-based non-price determinants of firewood and chips, and also related estimation problems. Focusing on hurdles in fuel choice and consumption decision, this chapter has identified stage-based socio-economic determining (or non-price) factors of selection of firewood and chips and its non-zero quantity consumption at household levels in rural India. The double hurdle model of Cragg (1971) provides significant factors like family composition, education level of household and its head, land assets, religion and social caste groups, female-headed household, etc. in the first hurdle of firewood and chips selection. Secondary and above education level of household head, female-headed household, religions and social caste groups, family size, land holding, average education and average age composition of household and income are significant determining factors in the second hurdle of quantity consumption decision. The results reflect that significant socio-economic variables in the first hurdle of fuel selection are different from the second hurdle or quantity choice of firewood and chips consumption.

This chapter provides a comparison between the inclusion and exclusion of land assets like land ownership (LO) in the said model. This study highlights the importance of land, which also reflects the asset structure of households in rural India. Firewood and chips, being a dirty fuel, has several ill effects, and such fuel consumption should be discouraged raising girl's education and attaining the targets of SDGs. Our results suggest to redistribute land assets, reduce family size and raise level of education for all, especially for female.

More specifically, policymakers should focus on policy redesign for improving female education beyond the secondary level. Empirical findings suggest shifting some firewood and chips consumers to other clean fuels through educational attainment of household head or raising awareness regarding dirty fuel consumption. Female participation should be encouraged at household-level decision-making that may help shifting household from dirty to clean fuels consumption. It is true that women suffer directly from ill effects of dirty fuel consumption and so their voice in decision-making of households might bring about some positive change in fuel consumption patterns of rural household in India.

Chapter 8
Firewood and Chips Consumption in Bottom and Top Quartile Groups

8.1 Introduction

The earlier chapter has identified income as one of several socio-economic determining factors of choice of firewood and chips in rural India. However, the previous chapter did not highlight the variation of determining factors due to income quartile groups in rural India. This study has attempted to explore further the determinants of consumption of firewood and chips in income quartile groups of households in rural India. Here, this study analyses the earlier model focusing on two extreme income groups such as the highest and lowest and income quartiles. Here, we have considered households in the bottom quartile (Q1) and the top quartile (Q2) in rural India. Does the hurdle of fuel choice and consumption decision vary across quartiles? The double hurdle model (DHM) is applied here for estimation purposes and understanding whether a change in economic status affects the firewood and chips consumption behaviour of households in rural parts of India.

8.2 Results and Discussions

The double hurdle model is used here for the said the National Sample Survey Office (NSSO) data set and results are reported in Table 8.1 and discussed as follows. Table 8.1 represents the estimated results of the participation and quantity equations of the double hurdle model for firewood and chips consumption decisions for the bottom and top quartile groups. The results indicate that income significantly determines firewood and chips as cooking fuel for bottom as well as top quartiles. However, the results of income are in the opposite direction: the income coefficient is positive in the bottom quartile while it is negative in the top quartile. It indicates

This chapter is partly based on Daripa and Dinda (2024).

S. Daripa, S. Dinda, *Hurdles in Fuel Choice in Rural India*, SpringerBriefs in Geography,
https://doi.org/10.1007/978-3-031-71605-8_8

Table 8.1 Quartile group-wise estimated results of double hurdle model

	Bottom quartile Group		Top quartile group	
Variables	Participation equation	Quantity equation	Participation equation	Quantity equation
Constant	−21.384*** (−3.43)	7.873* (1.95)	18.073*** (6.64)	12.54*** (5.83)
LogMPCE	6.62*** (3.47)	−1.27 (−1.03)	−4.132*** (−6.40)	−2.08*** (−4.05)
LogMPCE_sq	−0.498*** (−3.41)	0.113 (1.21)	0.235*** (6.14)	0.125*** (4.14)
LO	*−15.78**** *(−6.27)*	*−10.6**** *(−7.76)*	*−35.524**** *(−16.01)*	*−13.98**** *(−5.48)*
Hhsize	0.03*** (3.76)	0.04*** (9.86)	0.043*** (5.03)	0.064*** (9.91)
Age	0.007 (1.24)	0.015*** (4.71)	0.0214*** (3.85)	0.0034 (0.75)
Age_sq	−0.00002 (−0.31)	−0.0003*** (−7.37)	−0.0002*** (−3.68)	−0.0001** (−2.02)
Edu	0.024** (2.19)	−0.0015 (−0.24)	−0.106*** (−13.3)	−0.018** (−2.15)
Age_head	0.018** (2.68)	−0.006 (−1.51)	0.021*** (3.61)	0.01** (2.04)
Age_head_sq	−0.0002*** (−3.45)	0.0001** (2.49)	−0.0002*** (−3.38)	−0.0001 (−1.61)
Sex_head				
Female	0.06 (1.24)	−0.1*** (−3.69)	0.08** (2.11)	0.01 (0.34)
Edu_head				
Through EGS/NFEC/AEC	0.061 (0.28)	−0.04 (−0.32)	−0.3 (−1.29)	0.09 (0.47)
Through TLC	−0.004 (−0.01)	−0.106 (−0.51)	0.19 (0.38)	−0.31 (−0.98)
Others	−0.15 (−0.82)	−0.07 (−0.65)	−0.42 (−1.19)	−0.36 (−1.21)
Below primary	0.048 (1.02)	0.049** (2.04)	0.145** (2.63)	0.128*** (3.25)
Primary	−0.045 (−0.95)	0.012 (0.49)	0.143*** (2.77)	0.1** (2.62)
Middle	0.011 (0.21)	−0.06** (−2.25)	0.117** (2.35)	0.036 (0.96)
Secondary	−0.18*** (−2.99)	−0.08** (−2.33)	−0.088* (−1.75)	−0.096** (−2.44)
Higher secondary	−0.27*** (−3.22)	−0.095** (−1.98)	−0.192*** (−3.26)	−0.095** (−1.98)

(continued)

Table 8.1 (continued)

Variables	Bottom quartile Group Participation equation	Quantity equation	Top quartile group Participation equation	Quantity equation
Diploma/certificate course	−0.676* (−1.79)	−0.055 (−0.20)	−0.274*** (−3.06)	−0.34*** (−4.17)
Graduate	−0.673*** (−6.55)	0.0514 (0.77)	−0.305*** (−4.80)	−0.145** (−2.68)
Postgraduate and above	−0.52** (−2.43)	−0.03 (−0.22)	−0.383*** (−4.90)	−0.301*** (−4.13)
Social group				
ST	0.36*** (6.29)	0.098*** (3.66)	0.322*** (7.9)	0.361*** (10.32)
SC	0.064 (1.41)	−0.149*** (−5.83)	0.304*** (7.63)	0.043 (1.21)
OBC	0.063 (1.63)	−0.215*** (−9.71)	0.29*** (11.0)	−0.04 (−1.57)
Religion				
Islam	−0.223*** (−5.41)	0.099*** (4.07)	0.097** (2.50)	0.1214*** (4.03)
Christianity	−0.144* (−1.71)	0.21*** (5.49)	0.289*** (6.45)	0.294*** (8.12)
Others	0.28** (2.72)	0.249*** (4.87)	−0.407*** (−9.9)	−0.106** (−2.49)
No observations	14,914		14,917	
Sigma	0.8577		0.8904	
Covariance	−0.7757		0.0260	

Note: '***', '**' and '*' denote 1%, 5% and 10% level of significance, respectively.
Source: Daripa and Dinda (2024)

that the bottom quartile has a positive effect whereas the top quartile has a negative effect on the selection of firewood and chips in rural India. In other words, households in the bottom quartile raise their selection probability of dirty fuel like firewood and chips, whereas, households in the top quartile decline the probability of selecting firewood and chips and also reduce its consumption amount. Here, rural rich households reduce the consumption of firewood and chips in India. However, rural poor households raise their firewood and chips consumption in India.

Land is an important asset of rural households in India. The coefficient of land holding (LO) is negative and statistically significant in both equations for bottom and top quartiles. It has significant impact on the selection of firewood and chips and its consumption in rural households. The results reflect that land assets, LO, leads to reduction in the probability of selection of firewood and chips as well as decreased consumption amount of firewood and chips for both the lowest and highest quartile groups. Hence, it can be said that the results of land holding and income imply that the advancement of economic position of households has a negative impact on the selection as well as consumption of firewood and chips by rural households.

The impact of household or family size on firewood and chips consumption is positive in bottom and top quartiles. The coefficient of household size is positive in both the participation and quantity equations. So, the probability of selection and consumption amount of firewood and chips increases directly with household size irrespective of rich and poor groups. This result suggests that family size raises the choice and quantity demand for firewood and chips for cooking purposes irrespective of their income category.

Socio-economic variables are crucial determining factors in the choice of firewood and chips and consumption decision. Social caste (scheduled tribes (ST), scheduled castes (SC) and other backward class (OBC)) factors are important for the probable selection of firewood and chips for the rich or top quartile as it is significant in the participation equation of the top group, but its consumption amount is significantly increased only in scheduled tribes (ST) in the top quartile. ST is indifferent in the selection of firewood and chips and its consumption irrespective of rich and poor classes. In the bottom quartile, SC and OBC significantly reduce the consumption of firewood and chips in rural areas.

It is observed from the results that the average age of the household positively affects the amount of firewood and chips consumption of the bottom quartile; however, the results vary and are different for the top quartile, where the average age of households significantly determines the choice of firewood and chips but have no impact on its consumption amount. The age of the household head determines significantly both the choice and consumption of firewood and chips in the top quartile but does not have any impact on the amount of firewood and chips consumption in the bottom quartile.

The results indicate that probability of selection of firewood and chips is positive for female headed households for top quartile or rich households. However, for the lowest quartile or poor households, it is observed that female-headed households report a decline in the amount of consumption of firewood and chips.

The results reflect that for bottom quartile group, the average educational level of households only have significant positive impact on the selection of firewood and chips, whereas for the top quartile group, the average educational level of households have a negative impact both on the selection and on the amount of consumption of firewood and chips. Hence, it can be inferred that change in economic position of household alters the effect of average educational level of households in rural India. The impact of educational attainment of household head is also significant for both groups as household head is the major decision-maker of their family. The results suggest that there is a positive probability of selection of firewood and chips, and positive amount of its consumption is made at lower levels of educational attainment of household head, whereas if a minimum of at least secondary level of education is achieved, then the probability of choice and amount of consumption of firewood and chips comes down for both quartile (top and bottom) groups. Thus, unlike the effect of average educational level of household, the effects of educational level of the household head do not vary across quartile groups.

The results reveal that three social castes (SC, ST and OBC) have a positive probability of selection of firewood and chips for both groups. The quantity equation suggests that scheduled tribes (ST) households for the lowest quartile group make a positive amount of consumption of firewood and chips. Results of scheduled castes (SC) are interesting with contrast views. As per the participation equation, SC and OBC households of bottom quartile groups reduce the amount of consumption of firewood and chips, if it is selected, but it is insignificant in the participation equation. However, both SC and OBC households of the top quartile group have a positive probability of selection of firewood and chips, but it is insignificant in quantity equation. For the bottom quartile, both SC and OBC make a reduction in the amount of consumption of firewood and chips compared to general caste households. Hence, it can be inferred that the effect of different social groups on firewood and chips selection and consumption decision vary across quartile groups.

Quartile results reveal clearly that the economic position of households has strong effects on fuel consumption decision for firewood and chips. The effects of income are positive on the probability of selection of firewood and chips for the bottom quartile, while it is negative for the top quartile. Income determines a positive amount of consumption of firewood and chips for the top quartile group only. Land holdings have an indirect effect on selection and consumption amount. It is noted that the indirect effect is captured in the average educational level of the top quartile households. Educational levels of households might change their attitude towards dirty fuel like firewood and chips selection and consumption.

8.3 Conclusion

The results reflect significant socio-economic variables in the first hurdle of fuel selection are different from second hurdle or quantity choice of selected fuel. The significant socio-economic factors such as household composition, features of household head, land and castes at the bottom quartile raise the probability of firewood and chips selection, while the second hurdle shows a different set of factors with opposite signs. The results of quantity equation reveals that positive amount of consumption of firewood and chips is associated with income, household composition, ST, religious groups and primary level education of household head, while declining quantity consumption is associated with land, female headed household, SC, OBC, etc.

From the entire quartile results, the change in economic position has both direct and indirect effects on fuel consumption decision. The direct effect is reflected by the effect of household size, age and ST social group, whereas the indirect effect is reflected by land holdings, average education of households, etc. Hence, the advancement of the economic position of the households changes their attitude towards dirty fuel like firewood and chips selection as well as its consumption. Discrete policy choice will be more effective than uniform fuel policy. Recently, the

Government of India has adopted such policies like the Ujjwala gas scheme for the rural poor and more than 8 crore households benefited from the said programme. It should be noted that the policymakers should be given more focus on the accessibility and availability of alternative clean fuels for all sections of rural households especially old age and female-headed households. The policymakers may also think about educational attainment and rural land distribution. Asset reallocation policy should be assigned with carbon accounting and green credit that definitely ensures Sustainable Development Goal 7 (SDG 7).

Part IV
Analysis

Chapter 9
Firewood and Chips Consumption in the Largest Populated State in India: Uttar Pradesh

9.1 Introduction

For analysis purposes, this chapter highlights the socio-economic factors associated with firewood and chips consumption in Uttar Pradesh (UP), which is the largest state in terms of population in India (Census 2011). In this context, we examine the fractal class-wise income distribution of rural households by their primary sources of cooking fuels in Uttar Pradesh (see Table 9.1 for details). Table 9.1 shows that firewood and chips consumption is high at bottom fractal classes, but it declines with higher fractal classes; LPG consumption increases with income, but around 30–34% households use dung cake as cooking fuel irrespective of their income classes in UP. Table 9.1 also shows that firewood and chips consumption is the highest (around 56.1% rural households) in UP followed by dung cake (33.4% household size—Hhsize). This study reveals that the role of the economic condition of households is important in rural UP. These observations support literature and our presumption that Uttar Pradesh is a poor state but their firewood and chips consumption is high.

Social caste is a dominating factor in UP and that is also reflected in part B of Table 9.1, which shows the distribution of social castes by primary source of cooking fuels. It is clearly observed that 69.3% households of scheduled tribes (ST) consume firewood and chips, and this percentage decreases with social caste ladder. With these preliminary observations, we apply Cragg's double hurdle model in UP.

S. Daripa, S. Dinda, *Hurdles in Fuel Choice in Rural India*, SpringerBriefs in Geography,
https://doi.org/10.1007/978-3-031-71605-8_9

Table 9.1 Percentage distribution of households in the fractal class of income with primary source of cooking fuels in rural Uttar Pradesh

Fractal class of Income	Primary source of cooking fuels in rural India									
	Coke, coal	Firewood and Chips	LPG	Gobar gas	Dung cake	Charcoal	Kerosene	Electricity	Others	No cooking arrangement
0–5%	0	65.8	0	0	23.4	0	0	0	10.2	0.6
5–10%	0	70.2	0	0	26.7	0	0	0	3.2	0
10–20%	0.7	63.2	1.3	0	30.6	0	0	0	4.1	0
20–30%	0	61.6	0.7	0	34.7	0	0	0	3.0	0
30–40%	0.2	71.9	1.1	0	24.9	0	0	0	1.8	0
40–50%	0.3	64.1	2.0	0	29.2	0	0	0	4.4	0
50–60%	0	60.0	4.8	0	33.4	0	0.8	0	0.9	0
60–70%	0.2	52.2	6.8	0	38.1	0	0	0	2.8	0
70–80%	0.7	50.5	8.3	0	38.7	0	0	0	1.7	0.2
80–90%	0.3	51.0	10.9	0	36.3	0	0.2	0	1.3	0
90–95%	0	39.7	13.3	0	42.0	0	0.3	0	4.4	0.2
95–100%	0.2	28.6	28.3	0	34.3	0	0	0.5	0.5	7.5
Average	**0.2**	**56.1**	**6.7**	**0**	**33.4**	**0**	**0.1**	**0**	**2.8**	**0.6**
Average MPCE (₹)	1079.1	1045.7	1882.9		1195.1		1201.6	9846.4.5	989.2	4566.2
Part B										
Caste	Coke, coal	Firewood and chips	LPG	Gobar gas	Dung cake	Charcoal	Kerosene	Electricity	Others	No cooking arrangement
ST	0	69.3	9.6	0	20.8	0	0	0	0.4	0
SC	0	62.7	2.3	0	30.9	0	0.1	0	3.0	1.0
OBC	0.4	53.9	6.4	0	35.6	0	0	0.1	3.0	0.5
Others	0.1	51.4	14.6	0	30.9	0	0.5	0	2.0	0.6
Average	0.2	56.1	6.7	0	33.4	0	0.1	0	2.8	0.6

Source: NSSO 68th Round

9.2 Result and Discussion

The results of the double hurdle model are reported in Table 9.2. Two hurdles are reported in two equations, i.e. participation equation and quantity equation. The participation equation identifies determining factors for the selection of firewood and chips at the household level in rural Uttar Pradesh. The quantity equation points out the determining factors of actual firewood and chips consumption in rural areas of Uttar Pradesh.

It is observed from the participation equation that income and its square, land holding (LO), household size (Hhsize), average education level of the household denoted by edu, sex of the household head being female, education level of the household head at secondary, graduate and postgraduate levels, SC social group and all major religious groups are statistically significant and affect the choice of firewood and chips as the major cooking fuel in rural Uttar Pradesh.

The quantity equation reveals a different picture. It is seen that the household size, sex of the household head being female, different social groups and others religious group significantly determine the amount of firewood and chips consumption in rural Uttar Pradesh. The results of the double hurdle model are now discussed in detail.

It is seen that income denoted by log monthly per capita consumption expenditure (MPCE) significantly determine the choice of firewood and chips as the major cooking fuel in rural Uttar Pradesh, but it does not significantly determine the amount of firewood and chips consumption. The positive sign of the coefficient of log MPCE and negative sign of the coefficient of log MPCE_sq reflects that with an increase in income of the household, the probability to choose firewood and chips increases but at a diminishing rate.

The coefficient of LO reflects that with an increase in land holding of the rural households, the probability of selection of firewood and chips declines in rural Uttar Pradesh, but land holding does not significantly determine the amount of firewood and chips consumption in rural UP.

It is observed from the model that household size significantly determines the choice of firewood and chips as the major cooking fuel and its amount of consumption in rural Uttar Pradesh. The positive sign of the coefficients further reflects that an increase in household size increases the probability to choose and amount of firewood and chips consumption. This is possibly because an increase in household size leads to increased demand for firewood and chips as primary cooking fuel.

The effect of age of the household is reflected from two variables: variables pertaining to average age of the household and its square denoted by age and age_sq and the variables related to the age of the household head and its square denoted by age_head and age head_sq. It is seen from Table 9.2 that these are insignificant.

The effect of educational level of the household is reflected from two variables: variables related to average educational level of the household denoted by edu and variables related to the educational level of the household head, which is a categorical variable, and we have considered illiterate as the reference category. Table 9.2 shows that the average educational level of the household significantly determines

Table 9.2 Results of double hurdle model in Uttar Pradesh

	Participation equation		Quantity equation	
Variable name	Coefficient	t statistic	Coefficient	t statistic
Constant	−4.9187**	−2.03	2.593*	1.73
LogMPCE	1.9583***	3.00	0.415	1.01
LogMPCE_sq	−0.1534***	−3.48	−0.032	−1.12
LO	−12.738***	−2.97	−2.967	−1.39
Hhsize	0.0366***	3.30	0.043***	8.63
Age	0.002	0.20	0.006	1.30
Age_sq	0.00007	0.59	−0.0002***	−2.96
Edu	−0.0357**	−2.33	−0.013	−1.63
Age_head	−0.0005	−0.05	−0.0013	−0.27
Age_head_sq	−0.00004	−0.37	0.00003	0.62
Sex_head				
Female	−0.196**	−2.62	−0.1265***	−3.24
Edu_head				
Through EGS/NFEC/AEC	−0.384	−1.17	0.1858	1.00
Through TLC	−1.146	−1.34	−1.272	−1.72
Others	0.633	1.33	−0.072	−0.42
Below primary	−0.11	−1.27	0.0318	0.75
Primary	−0.072	−0.91	−0.0495	−1.32
Middle	0.0596	0.76	−0.0218	−0.60
Secondary	−0.286***	−3.41	−0.0294	−0.67
Higher secondary	−0.1502	−1.47	−0.0163	−0.32
Diploma/certificate course	−0.6243	−1.47	0.5714*	1.88
Graduate	−0.4116***	−3.62	−0.0291	−0.45
Postgraduate and above	−0.5889***	−3.93	−0.1382	−1.43
Social group				
ST	−0.0512	−0.25	0.17516*	1.70
SC	0.13126*	1.84	−0.0724**	−2.06
OBC	0.07217	1.30	−0.0485*	−1.66
Religion				
Islam	−0.329***	−5.42	0.007812	0.23
Christianity	−1.176**	−2.42	0.611073	1.42
Others	−0.756*	−1.67	−0.64784*	−1.95
No. of observations	5915			
Sigma	0.7374129			
Covariance	0.0355825			

Note: '***', '**' and '*' denote 1%, 5% and 10% level of significance, respectively.
Source: Author

the probability of choice of firewood and chips as the major cooking fuel, but not its quantity. The negative sign of the coefficient further states that with an increase in the average educational level of the household, the probability of choice of firewood and chips declines in rural Uttar Pradesh. Considering the educational level of the household head, it is observed that the educational level of the household head at secondary, graduate and postgraduate level significantly determine the choice of firewood and chips as the major cooking fuel in rural Uttar Pradesh, and the negative sign of their coefficients reflects that an increase in the educational level of household head lowers the probability of selection of firewood and chips as the major cooking fuel compared to the illiterate households. In the case of quantity equation, it is observed that only a diploma or certificate course is weakly significant for determining the amount of firewood and chips consumption.

It is further observed from the model that compared to male-headed households, female-headed households show a decline in the probability to choose firewood and chips as the major cooking fuel as well as report a decline in the amount of firewood and chips consumption in rural Uttar Pradesh. This might be possibly because women are the direct sufferers of the ill effects of firewood and chips consumption.

Considering the effect of different social groups on firewood and chips choice and consumption, it is seen that compared to general caste households, SC households reveal a positive probability to choose firewood and chips, but the amount of firewood and chips consumption declines compared to general caste households, if it is selected. Other backward class (OBC) households also show a decline in the amount of firewood and chips consumption compared to general caste households, but firewood and chips consumption of scheduled tribes (ST) households increased compared to general caste households.

The effect of different religious groups on firewood and chips choice and consumption is also reflected in Table 9.2. Compared to the base category, i.e. Hindu households, all three major religious groups show a decline in the probability to choose firewood and chips as the major cooking fuel, and in the case of the amount of consumption of firewood and chips, only the other religious group is significant and that too has a negative impact on the amount of firewood and chips consumption compared to Hindu households.

9.3 Conclusion

This chapter observes that the socio-economic and demographic factors crucially determine cooking fuel consumption decision of rural households of India's largest populated state, Uttar Pradesh. Socio-economic factors like income, land, family size, education and female-headed households are the most significant determining factors of firewood and chips selection in rural UP. However, family size, female-headed households and social castes determine the actual consumption of firewood and chips in rural areas of Uttar Pradesh. Family size and female-headed households are the most significant determining factors of firewood and chips selection and consumption, but their effects are in opposite directions, i.e. contradict each other. It is good that female-headed households reduce firewood and chips consumption as well as their choice decisions.

Chapter 10
Choice of Cooking Fuel Across Castes in Rural India

10.1 Introduction

This study deals with understanding the regional, demographic and socio-economic factors that determine the choice of cooking fuel of households in rural India. This study also looks into the caste disparity in the context of the use of cooking fuels. This chapter has attempted to provide a comprehensive discussion on the choice of cooking fuel across social castes in rural India. Literature highlights the fuel consumption issues and related cooking fuel use in India; however, no study has focused on the fuel consumption pattern of social groups or determining factors of choice of cooking fuels across castes in rural India.

This chapter is organised as follows: Sect. 10.2 describes the cooking fuel consumption pattern across social castes in rural India; Sect. 10.3 provides state-wise distribution of castes in India; Sect. 10.4 discusses the distribution of rural households of socially deprived caste by sources of firewood and chips; Sect. 10.5 identifies the determining factors of using double hurdle model (DHM) analysis; and finally, Sect. 10.6 concludes with remarks.

10.2 Cooking Fuel Consumption Pattern Across Social Caste in Rural India

Considering the data on social groups, we find four major social castes existing in India. These are scheduled tribes (ST), scheduled castes (SC), other backward classes (OBC) and others (it is also termed as general caste). The lower part of Table 3.3 provides the frequency distribution of households classifying the above-said social castes, and their representations are ST (16.64%), SC (17.08%), Others (26.682%) and OBC (39.46%).

S. Daripa, S. Dinda, *Hurdles in Fuel Choice in Rural India*, SpringerBriefs in Geography,
https://doi.org/10.1007/978-3-031-71605-8_10

Now, this study investigates the cooking fuel choice among ten primary sources of fuels. Households of each social caste are distributed among ten primary cooking fuels. Table 10.1 provides the details of representation of each social caste as per their primary sources of cooking fuels in rural India. Table 10.1 shows that firewood and chips is the main source of cooking fuel of all castes such as scheduled castes (SC), scheduled tribes (ST), other backward classes (OBC) and others (or general) caste. Firewood and chips is the main source of cooking fuel of ST (87%), SC (69.8%), OBC (66.4%) and others (57%). The percentage share of firewood and chips of all castes is arranged in descending order starting from ST. Tribal households (87%) are highly dependent on firewood and chips for their cooking fuel, and least dependent is others or general caste (57%).

It is also noted that 5.3% of scheduled tribes (ST) households use LPG. Similarly, 8.9% of scheduled castes (SC), 16% of other backward classes (OBC) and 23.3% of others or general caste households use LPG for cooking purposes. Dung cake is used the least around 1.4% of ST and a maximum of 12.2% of SC households. SC and OBC households use dung cake more compared to the rest. Kerosene is used least in ST (0.5%) households and maximum of SC (1.1%) households. 'no cooking arrangement' is an important source for all castes. Here, 3.1% of ST households have 'no cooking arrangement'. It is quite high in this study, and it should have certain implications and needs to be explored later. Next, we examine the distribution of social caste across regions or/and states of India. Is the state preference dominated by caste? Next, this section attempts to explore it.

10.3 Distribution of Castes Across States in India

The distribution of these social castes across states of India is presented in Table 10.2 in terms of percentage share of each caste in respective states. Table 10.2 shows state-wise distribution of castes, and it also displays the widespread social groups across rural India.

Table 10.2 provides the ascending order of Indian states as per percentage share of each caste. The highest concentration of scheduled tribes (ST) is observed in the Northeastern states of India (such as Arunachal Pradesh, Meghalaya, Nagaland, Manipur, Mizoram, Tripura and Assam) followed by central India (like Madhya Pradesh, Orissa, Jharkhand and Chhattisgarh). Regional dominance of scheduled tribes (ST) is clearly observed in Table 10.2. The scheduled caste (SC) concentration is the highest in Uttar Pradesh followed by West Bengal, Andhra Pradesh, Punjab, Tamil Nadu, Orissa, Bihar, Rajasthan, etc. The highest OBC concentration is observed in Uttar Pradesh followed by Tamil Nadu, Andhra Pradesh, Bihar, Maharashtra, Kerala, etc. Others (general castes) are maximum in West Bengal, followed by Maharashtra, Jammu & Kashmir, Uttar Pradesh, Assam, etc.

Table 10.1 Percentage of households of each caste with primary source of energy for cooking in rural India

Social group	Coke and coal	Firewood and chips	LPG	Gobar gas	Dung cake	Char coal	Kerosene	Electricity	Others	No cooking arrangement
ST	1.1	**87.0**	5.3	0	1.4	0	0.5	0	1.6	3.1
SC	1.1	**69.8**	8.9	0	12.2	0	1.1	0	6.0	0.9
OBC	0.9	**66.4**	16.0	0.2	11.2	0	0.9	0.1	3.6	0.8
Others	1.7	**57.0**	23.3	0.5	8.3	0	0.8	0.1	6.7	1.5

Source: 68th Round NSSO Report

Table 10.2 State-wise distribution of social caste groups across rural India

State	ST	State	SC	State	OBC	State	others
Arunachal Pradesh	8.26	Uttar Pradesh	14.72	Uttar Pradesh	13.45	West Bengal	12.04
Meghalaya	8.13	West Bengal	10.44	Tamil Nadu	10.67	Maharashtra	9.26
Nagaland	6.71	Andhra Pradesh	7.69	Andhra Pradesh	8.6	Jammu & Kashmir	8.51
Madhya Pradesh	6.41	Punjab	6.79	Bihar	8.4	Uttar Pradesh	7.37
Orissa	6.4	Tamil Nadu	6.76	Maharashtra	6.87	Assam	6.99
Manipur	6.21	Orissa	5.59	Kerala	6.67	Andhra Pradesh	5.2
Mizoram	6.21	Bihar	5.42	Madhya Pradesh	5.18	Himachal Pradesh	5.09
Tripura	5.58	Rajasthan	5.13	Rajasthan	5.11	Kerala	4.84
Assam	5.04	Maharashtra	4.71	Orissa	4.92	Bihar	4.57
Jharkhand	5.03	Madhya Pradesh	4.18	Karnataka	4.88	Punjab	4.24
Chhattisgarh	4.88	Haryana	4.02	Jharkhand	3.38	Uttaranchal	4.15
Maharashtra	4.54	Himachal Pradesh	3.38	Gujarat	3.27	Haryana	3.85
Gujarat	4.42	Karnataka	3.06	Assam	3.05	Orissa	3.81
Rajasthan	4.27	Assam	2.66	Chhattisgarh	2.83	Karnataka	2.92
Andhra Pradesh	2.86	Jharkhand	2.49	Manipur	2.58	Madhya Pradesh	2.81
Sikkim	2.6	Kerala	2.27	Haryana	1.6	Rajasthan	2.67
Jammu & Kashmir	2.33	Tripura	2.24	West Bengal	1.59	Gujarat	2.31
Himachal Pradesh	2.02	Uttaranchal	2.01	Himachal Pradesh	1.26	Tripura	2.04
West Bengal	2.01	Chhattisgarh	1.89	Sikkim	1.25	Arunachal Pradesh	1.32
Karnataka	1.21	Jammu & Kashmir	1.67	Jammu & Kashmir	1.14	Jharkhand	1.31
D & N Haveli	0.82	Gujarat	1.3	Tripura	0.87	A & N Islands	1.12
Uttar Pradesh	0.67	Manipur	0.46	Punjab	0.76	Goa	0.74
Lakshadweep	0.61	Sikkim	0.35	Uttaranchal	0.59	Manipur	0.65
Tamil Nadu	0.55	Pondicherry	0.26	Pondicherry	0.4	Chhattisgarh	0.6
A & N Islands	0.5	Arunachal Pradesh	0.19	A&N Islands	0.2	Tamil Nadu	0.39
Bihar	0.44	Delhi	0.13	Daman & Diu	0.12	Meghalaya	0.29
Uttaranchal	0.39	Daman & Diu	0.07	Goa	0.1	Chandigarh	0.28
Kerala	0.31	Chandigarh	0.04	Chandigarh	0.06	Delhi	0.22
Haryana	0.21	Goa	0.03	Arunachal Pradesh	0.06	Sikkim	0.12

(continued)

Table 10.2 (continued)

State	ST	State	SC	State	OBC	State	others
Goa	0.15	Mizoram	0.02	Delhi	0.05	Daman & Diu	0.11
Daman & Diu	0.11	D & N Haveli	0.02	Mizoram	0.04	Mizoram	0.07
Punjab	0.03	Meghalaya	0.01	Nagaland	0.03	D & N Haveli	0.05
Delhi	0.03	Nagaland	0	D & N Haveli	0.02	Pondicherry	0.04
Chandigarh	0.02	Lakshadweep	0	Meghalaya	0.01	Lakshadweep	0.01
Pondicherry	0.01	A&N Islands	0	Lakshadweep	0	Nagaland	0

Source: Author

Considering the fuel choices of ST concentrated states, we further observed that Meghalaya and Tripura consider firewood and chips as the most preferred fuel, while others like Nagaland, Manipur, Mizoram, Arunachal Pradesh, etc. consider LPG as the most preferred cooking fuel. We need to explore the reason behind it.

10.4 Distribution of Rural Households of Socially Deprived Castes with Sources of Firewoods and Chips

Following Table 10.1, the choice of firewood and chips is the prime cooking fuel for socially deprived castes (ST and SC) and also true for other backward classes (OBC). More than 60% of them (SC, ST and OBC) select firewood and chips for their cooking fuel. Now the question of the sources of firewood and chips arises. Table 10.3 provides the distribution of households by sources of firewood and chips of socially deprived castes (ST, SC and OBC).

Major sources of firewood and chips are only purchase, only home-grown stock, both purchase and home-grown and only free collections. Only free collection is identified as the highest or top source of firewood and chips among ST and SC; 45.25% of ST, 41.92% of SC and 31.55% of OBC households of socially deprived groups get their cooking fuel from only free collection sources. The second highest source is only home grown for ST (25.45%), and their third source is only purchase (21.2%). However, for SC, the second and third highest sources are, respectively, only purchase (32.37%) and only home grown (17.79%). For OBC, the first, second and third sources of firewood and chips are, respectively, only purchase (32.59), only free collection (31.55) and only home-grown stock (28.16). Fig. 10.1 displays these in multiple bars diagram with a clear comparison among ST, SC and OBC. We observe that the first, second and third ranks (in terms of percentage) in the case of only free collections are held by ST (45.25%), SC (41.92%) and OBC (31.55%), respectively. After only free collections, SC and OBC are dominated in the source of only purchase. OBC and ST dominate in only home-grown source.

Majority of ST, SC and OBC have free collection source of firewood and chips because they are deprived economically and socially which is reflected in their land

Table 10.3 Firewood and chips source-wise distribution of rural households of deprived castes in India

	SC		ST		OBC	
	Frequency	Percentage	Frequency	Percentage	Frequency	Percentage
Only purchase	2787	32.37	1836	21.2	6350	32.59
Only home-grown stock	1532	17.79	2203	25.45	5488	28.16
Both purchase and home-grown	271	3.15	223	2.58	674	3.46
Only free collections	3609	41.92	3916	45.25	6149	31.55
Only exchange	22	0.26	7	0.08	48	0.25
Only gifts/charities	13	0.15	7	0.08	43	0.22
Others	376	4.37	464	5.36	735	3.77
Total	8610		8655		19,487	

Source: Author

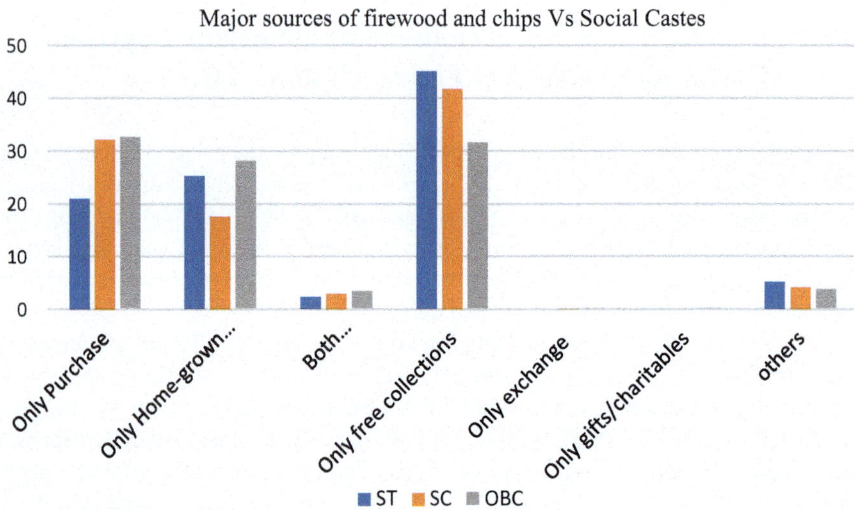

Source: Author

Fig. 10.1 Major sources of firewood and chips used by deprived castes in rural India. (Source: Author)

assets. Table 10.4 shows the distribution of land owned across socially deprived castes in rural India. Majority of the tribals (around 70.4% ST) have no land in rural parts of India. The highest landless households are ST (70.4%), followed by OBC (55.2% and SC (38.4%). Majority of deprived castes have 0.011 hectare land and 49.32% SC holds maximum among them (see Fig. 10.2).

Table 10.4 Distribution of land owned (LO) across castes in rural India

	ST		SC		OBC	
Land owned (Hectare)	Frequency	Percentage	Frequency	Percentage	Frequency	Percentage
0	6988	70.37	3910	38.36	12,996	55.19
0.001	6	0.06	7	0.07	5	0.02
0.002	332	3.34	417	4.09	1054	4.48
0.011	2034	20.48	5027	49.32	7817	33.20
0.012	550	5.54	811	7.96	1600	6.80
0.013	20	0.20	21	0.21	74	0.31
Total	9930	100.0	10,193	100.0	23,546	100.0

Source: Author

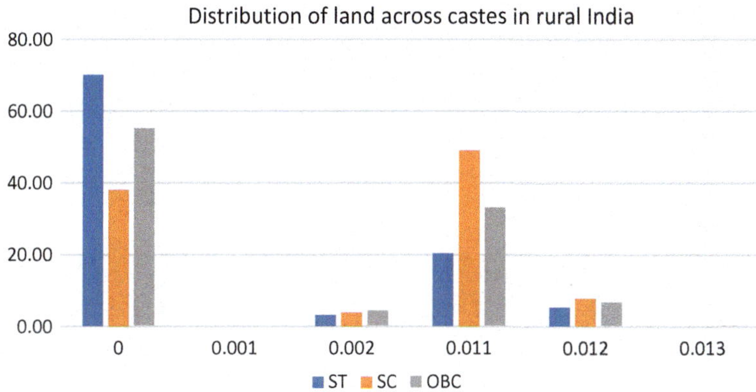

Source: Author

Fig. 10.2 Distribution of land (LO) across deprived castes in rural India. (Source: Author)

10.5 Determinants of Cooking Fuel Among Deprived Social Castes in Rural India

Desired and actual consumption are different due to certain factors which are associated with demographical factors and social conditions apart from economic situations. Missing information arises due to the gap between actual and desired level of consumption. In this context, we apply Cragg's double hurdle model for identifying the determinants of the choice of firewood and chips for socially deprived groups, here ST, SC and OBC. We have considered factors like income, land, family composition, characteristics (age, sex and educational level) of head of households, religious beliefs, etc. Following literature, monthly per capita consumption expenditure (MPCE) is used as proxy of income for the purpose of this study.

Table 10.5 provides brief and precisely focused results of two stage–based cooking fuel choice and consumption decision of three different social castes, namely ST, SC and OBC. The coefficients of income and its square, land, household size,

Table 10.5 Determinants of firewood and chips consumption of scheduled castes (SC) and scheduled tribes (ST) in rural India

	SC		ST		OBC	
Variable name	Participation equation	Quantity equation	Participation equation	Quantity equation	Participation equation	Quantity equation
Constant	−5.618** (2.76)	−4.952*** (−3.73)	8.0072*** (3.56)	2.224** (2.08)	3.6984*** (3.61)	1.487** (2.02)
LogMPCE	2.013*** (3.68)	2.39*** (6.59)	−1.601*** (−2.68)	0.5283 (1.8)	−0.5516** (−2.08)	0.58*** (2.98)
LogMPCE_sq	−0.1567*** (−4.25)	−0.159*** (−6.36)	0.0667 (1.69)	−0.0331 (−1.64)	0.0209 (1.21)	−0.0287** (−2.22)
LO	−21.447*** (−6.93)	−18.58*** (−10.82)	−36.02*** (−9.91)	−11.106*** (−6.02)	−19.76*** (−10.87)	−5.35*** (−4.30)
Hhsize	0.02516** (2.19)	0.0573*** (9.68)	0.0249 (1.84)	0.0695*** (11.85)	0.0327*** (5.24)	0.052*** (13.66)
Age	0.0221** (3.0)	0.012** (2.97)	0.012 (1.32)	−0.00034 (−0.08)	0.009** (2.13)	0.014*** (4.78)
Age_sq	−0.00026** (−2.85)	−0.0002*** (−4.70)	−0.00009 (−0.77)	−0.0001** (−2.06)	−0.0001 (−1.88)	−0.0003*** (−7.22)
Edu	−0.074*** (−6.12)	0.004 (0.56)	−0.0915*** (−6.37)	−0.0237*** (−3.28)	−0.0552*** (−7.55)	0.0157*** (3.013)
Age_head	0.02085** (2.51)	0.0034 (0.70)	0.0834*** (9.81)	0.011** (2.18)	0.011** (2.19)	−0.004 (−1.23)
Age_head_sq	−0.00024** (−3.0)	−3.06e-06 (−0.06)	−0.0009*** (10.95)	−0.00007 (−1.32)	−0.0001** (−1.98)	0.00006 (1.74)
Sex_head						
Female	0.08485 (1.54)	−0.06** (−2.01)	−0.1463** (−2.35)	−0.112*** (−3.58)	0.0647** (1.97)	−0.0227 (−1.03)
Edu_head						
Through EGS/NFEC/AEC	0.0635 (0.20)	0.1235 (0.79)	−0.094 (−0.33)	−0.123 (−0.95)	0.01 (0.05)	0.12 (1.04)

Through TLC	3.5528 (0.03)	0.4197 (1.04)	3.603 (0.03)	0.1287 (0.63)	0.618 (1.23)	−0.642** (−2.40)
Others	0.05958 (0.20)	−0.223 (−1.48)	0.272 (0.61)	0.0102 (0.07)	−0.435*** (−2.62)	0.129 (1.01)
Below primary	0.2336*** (3.79)	0.0847** (2.78)	0.0188 (0.25)	0.112*** (3.60)	0.115*** (3.08)	0.043 (1.86)
Primary	0.0261 (0.45)	0.07** (2.23)	0.0725 (0.95)	0.07097** (2.24)	0.097*** (2.58)	0.0593** (2.49)
Middle	0.024 (0.41)	0.0135 (0.41)	−0.0774 (−1.02)	0.0185 (0.55)	0.119*** (3.17)	−0.02 (−0.83)
Secondary	−0.11 (−1.62)	−0.0952** (−2.38)	−0.197** (−2.37)	−0.0645 (−1.62)	−0.17*** (−4.25)	−0.0546** (−1.99)
Higher secondary	−0.156 (−1.62)	−0.17*** (−3.28)	−0.279*** (−2.94)	−0.0293 (−0.61)	−0.285*** (−5.77)	−0.007 (−0.21)
Diploma/certificate course	−0.52** (−2.76)	−0.47** (−3.12)	−0.342 (−1.84)	−0.436*** (−3.65)	−0.38*** (−4.12)	−0.059 (−0.77)
Graduate	−0.467*** (−4.85)	−0.2008** (−2.98)	−0.2377** (−2.30)	−0.035 (−0.66)	−0.465*** (−8.17)	−0.028 (−0.64)
Postgraduate and above	−0.5235*** (−3.58)	−0.285** (−2.55)	−0.238 (−1.50)	−0.092 (−0.94)	−0.57*** (−7.10)	−0.048 (−0.71)
Religion						
Islam	−0.3684** (−2.24)	−0.0512 (−0.51)	0.0224 (0.21)	−0.0437 (−0.85)	0.0309 (1.03)	0.041** (2.08)
Christianity	−0.1647 (−1.10)	−0.025 (−0.28)	−0.0502 (−1.13)	0.262*** (12.36)	0.124 (1.65)	−0.0547 (−1.07)

(continued)

Table 10.5 (continued)

	SC		ST		OBC	
Variable name	Participation equation	Quantity equation	Participation equation	Quantity equation	Participation equation	Quantity equation
Others	−0.6972*** (−12.65)	−0.1286** (−2.71)	−0.3677*** (−6.41)	0.05142 (1.65)	−0.403*** (−5.20)	0.24*** (3.76)
No. of observations	10,192		9927		23,542	
Sigma	0.80827		0.7878893		0.94178	
Covariance	0.007956		0.0003943		−0.7957866	

Note: '***', '**' and '*' denote 1%, 5% and 10% level of significance, respectively.
Source: Author

age and its square, average education of family, age of the household head and its square, below primary, diploma, graduation and above level of education of household head, Islam and others group are statistically significant in the first stage of participation of SC group. In the second stage, significant coefficients of variables in quantity consumption of firewood and chips of SC group are income and its square, land, household size, age and its square, female-headed household, below primary, secondary and above level of education of household head and others group. Thus, significant factors for SC group are income, land, household size, average age of family, below primary and secondary and above education level of household for choosing firewood and chips and its consumption incidence. Hence, choice of firewood and chips consumption of scheduled castes (SC) depends on income, land, family composition including age and family size, education and female household head.

For scheduled tribes (ST), the coefficients of income, land, average education of family, age of household head, female-headed household, secondary and above education of household head and others group are statistically significant in participation decision stage. However, in the second stage, in quantity consumption decision of ST, only significant coefficients are land, average education of family, age of household head and female-headed household, and also added factors like household size, below primary, primary and diploma/certificate level of education of household head and Christian religion. Only land, education and female-headed household are significant factors for ST for the selection and consumption of firewood and chips. So, the choice of firewood and chips consumption of scheduled tribes (ST) depends on land, average family education and female-headed households.

For other backward classes (OBC), the coefficients of income, land, family size, average age and average education of family, age of household head and its square, female-headed household, below primary and above level of education of household head and others group are statistically significant in the first stage of participation decision of OBC group. In the second stage of quantity decision, significant coefficients of variables in quantity consumption of firewood and chips of OBC group are income and its square, land, family size, average level of education and average age of family members and its square, primary and secondary level of education of household head, Islam and others group. Here, the significant factors for OBC group are income, land, household size, average age and education level of family, primary and secondary education level of household head and others group for selecting firewood and chips and its consumption incidence. So, the choice of firewood and chips consumption of other backward classes (OBC) depends on income, land, family composition including age, family size and level of education, female-headed household, primary and secondary educational level of head of households and others religious group.

The coefficients of land are negative in both stages of decision that are taken by households of all castes. So, land-rich households of all castes reduce their consumption of firewood and chips. The sign of income varies from positive to negative across social castes. Income is an important decision-making factor for SC and

OBC, but not for ST. Household composition is also a significant deciding factor for SC and OBC, but not for ST. Female household head of ST significantly reduce consumption decision for firewood and chips. Less education level of household heads raises firewood and chips consumption in SC and OBC castes, while higher (graduation and above) education of heads of households reduces it in SC. Religious factor of others reduces firewood and chips consumption in SC and OBC caste groups.

10.6 Conclusion

This chapter has discussed the choice of cooking fuel across castes in rural India. The study has identified various demographic characteristics such as family size, average age of family, age and sex of family head; economic factors like income and land; and social factors like level of education of family as well as head of the family and religious beliefs,. which are highly associated with consumption decision for firewood and chips across castes (ST, SC and OBC) in rural India. Land-owned households of all castes reduce their consumption of firewood and chips. Land-owned rights might ensure a reduction of firewood and chips consumption. Hence, the government should redesign the land distribution policy for the deprived social castes across rural India. Female-headed household also reduces consumption decision for firewood and chips. The next chapter focuses on it and highlights the hurdles of female-headed households for cooking fuel choice and its consumption in rural India.

Chapter 11
Hurdles of Female-Headed Households for Cooking Fuel Choice and Consumption in Rural India

11.1 Introduction

Actual cooking fuel consumption decision at unit (household) level is important to attain Sustainable Development Goals (SDG) in 2030. It is an important research issue to identify the hurdles of clean fuel consumption and its access to rural India for attaining SDG7 in 2030. Who will make the cooking fuel choice and consumption decision in households in rural parts of India? Who will execute fuel consumption decision in rural India? What is the role of females in attaining SDGs within 2030? In reality, the household head generally takes the said decisions and female members execute them; however, it may vary from region to region. Now, consider the situation where female-headed households are engaged in decision-making as well as execution in rural parts of India. The least attention has been given to hurdles in fuel choice and quantity of fuel consumption decision of female-headed household levels in developing countries like India. Fuel is mostly required for cooking purposes in rural area. Hence, our focus is on female-headed households and their cooking fuel choice in rural India. Literature discusses fuel prices in the market system, while little attention has been given to non-price issues associated with the fuel choice decision of female-headed households in rural India.

This chapter investigates the hurdles of female-headed households in the selection of cooking fuel and its consumption intensity in rural India. This study examines and identifies stage-based determinants of cooking fuel choice and consumption decision of female-headed households in rural areas focusing on non-price factors. Using the double hurdle model (Cragg, 1971) this chapter attempts to identify the determining non-price factors at each stage of fuel consumption decision of female-headed household in rural India.

The rest of the chapter is organised as follows: Sect. 11.2 briefly describes fuel consumption patterns in female-headed households and its related basic

S. Daripa, S. Dinda, *Hurdles in Fuel Choice in Rural India*, SpringerBriefs in Geography,
https://doi.org/10.1007/978-3-031-71605-8_11

socio-economic information. Section 11.3 provides results and discusses the major findings with appropriate interpretations. Finally, Sect. 11.4 concludes with remarks.

11.2 Fuel Consumption Pattern in Female-Headed Households

The analysis of this study is mainly based on the last published quinquennial round, the 68th round[1] of National Sample Survey Office (NSSO) data (2011–2012), which was published by the Government of India in 2015. It is observed that around 10.69% household is female headed in rural India (Table 3.3). NSSO data provides individual's options of ten major alternative fuels for cooking purposes, namely firewood and chip, dung cakes, coke or coal, charcoal, kerosene, gobar gas, LPG, electricity, others and no cooking arrangements. NSSO also provides monthly per capita consumption expenditure (MPCE) which is taken as a proxy variable for income following literature.[2] Table 11.1 shows the distribution of cooking fuels of female-headed households in rural India.

Table 11.1 describes the distribution of cooking fuel types used in female-headed households and corresponding their mean income; 24.60% of female-headed rural

Table 11.1 Distribution of cooking fuel used in female-headed households in rural India and their mean income

Cooking code	Frequency	Percentage	Mean MPCE (₹)	SD MPCE	Max–Min
Coke and coal	53	0.83	1441.7	799.06	4707.5–524.3
Dung cake	361	5.66	1546.48	920.897	6629.4–322.0
Firewood and chips	***4091***	***64.13***	***1661.14***	***1338.27***	***9846.4–804.6***
Kerosene	85	1.33	2420.34	1558.2	9497.5–544.4
LPG	1569	***24.60***	2953.4	2656.77	56960.1–670.4
Electricity	9	0.14	3137.09	2744.69	9846.4–804.6
Gobar gas	*3*	*0.05*	*3078.57*	*3210.49*	*6763.9–888.4*
Charcoal	*1*	*0.02*	*1976.5*	–	*1976.5–1976.5*
Others	154	2.41	1179.5	657.13	5678.2–284.7
No cooking arrangement	*53*	*0.83*	*3138.29*	*2974.91*	*12453.1–150.1*
Total	6379	100	1984.24	1845.81	56960.1–150.1

Note: Sample weights are considered here.
Source: NSSO 68th Round

[1] It (68th round) was the last published quinquennial round NSSO data. The programme of quinquennial survey on consumer expenditure and employment & unemployment, adopted by the National Sample Survey Office (NSSO) since 1972–73.

[2] See Swarup and Rao (2015), Gundimeda and Kohlin (2008), Gupta and Kohlin (2006), Daripa and Dinda (2022), etc.

households use LPG as cooking fuel having their average income around ₹2953.4. However, 64.13% of female-headed rural households use firewood and chips as cooking fuel having an average income of around ₹1661.14 (Table 11.1). It is observed that firewood and chips (64.13%) is the most preferred cooking fuel at female-headed rural households in India, and follow up fuels are, respectively, LPG (24.60%), dung cake (5.66%), others (2.41%) and kerosene (1.33%). The most preferred cooking fuel of rural female-headed households is firewood and chips, and we need to know its sources. Table 11.2 provides the sources of firewood and chips that are used by female-headed households in rural India. The main source of firewood and chips is the free collection. Around 36.53% of female-headed households use free collection sources of firewood and chips, and the second highest source is purchase (30.19%). The third highest source of firewood and chips is home-grown stock (23.60%) that might include agricultural residuals.

In this context, we have also considered the land-holding pattern of female-headed households.

Table 11.3 provides the distribution of land ownership of female-headed households in rural India. The distribution of land of female-headed households is highly bi-modal in rural India.

Variation of land amount is limited, minimum no land (or 0 hectare) and maximum land amount of 0.013 hectare. So, land ownership of female-headed households is distributed within the narrow range of 0–0.013 hectare in rural areas. Table 11.3 shows that the majority of rural female-headed households (around 42.25%) have no land ownership. So, 2695 out of 6379 female-headed households are landless and only 23 female households have the maximum amount of land holding (around 0.013 hectare). However, 41.28% female-headed households have an amount of 0.011 hectare in rural India. Table 11.3 shows that the majority (42.25%) of them are landless and their sources of firewood and chips are mostly free collection or/and purchase.

Cooking fuel choice and consumption of female-headed households in rural India mostly depend on demographic, social and economic condition–related variables which are beyond market mechanism. Possibly, demographic and socio-economic factors govern the fuel choice of female-headed households. Table 11.4

Table 11.2 Sources of cooking fuel of female-headed households in rural India

Source of cooking fuels	Frequency	Percentage
Only purchase	1562	30.19
Only home-grown stock	*1221*	*23.60*
Both purchase and home-grown stock	185	3.58
Only free collection	*1890*	***36.53***
Only exchange of goods and services	14	0.27
Only gifts/charities	*18*	*0.35*
Others	284	5.49
Total	5174	100

Source: Author

Table 11.3 Land distribution of female-headed households in rural India

Land owned (Hectare)	Frequency	Percent	Cumulative
0.0	2695	42.25	42.25
0.0–0.001	2	0.03	42.28
0.001–0.002	497	7.79	50.07
0.002–0.011	2633	41.28	91.35
0.011–0.012	529	8.29	99.64
More than (>) 0.012	23	0.36	100.00
Total	6379	100.00	

Source: Author

Table 11.4 Distribution of social factors of female-headed households in rural India

Name of the variable	Category	Frequency	Percentage
Social groups	General	1646	25.81
	ST	958	15.02
	SC	1136	17.81
	OBC	2637	41.35
Religion	Hinduism	4821	75.59
	Islam	809	12.68
	Christianity	470	7.37
	Others	278	4.36
Educational level of the head of household	Illiterate	3538	55.47
	Through EGS/NFEC/AEC	17	0.27
	Through TLC	9	0.14
	Others	19	0.30
	Below primary	676	10.60
	Primary	682	10.69
	Middle	637	9.99
	Secondary	409	6.41
	Higher secondary	192	3.01
	Graduate	123	1.93
	Post-graduate and above	43	0.67

Source: Author

describes the distribution of social factors of female-headed households in rural India. The majority of households belong to category of OBC (41.35%) followed by others (25.81%), SC (17.81%) and ST (15.02%). Majority belongs to Hindu category (75.59%) followed by Islam (12.68%), Christianity (7.37%) and all the other religions (4.36%). Education levels of household heads are categorised (Table 11.4) as illiterate (55.47%), below primary (10.60%), primary (10.69%), middle (9.99%), secondary (6.41%) and highly educated (i.e. graduate and above (2.60%)). We have also considered demographic variables like family size, age composition of family and age and sex of the household head.

Table 11.5 State-wise distribution of female-headed households (%) and their average education level (year of schooling) in rural India

State	% of female-headed households in rural India	Mean educational level of female-headed households (years of schooling)
Jammu & Kashmir	2.43	3.56
Himachal Pradesh	5.28	4.86
Punjab	2.59	4.04
Chandigarh	0.06	4.32
Uttaranchal	2.4	3.37
Haryana	1.87	2.37
Delhi	0.05	10.48
Rajasthan	4.01	1.96
Uttar Pradesh	8.75	2.30
Bihar	3.89	2.41
Sikkim	1.19	4.20
Arunachal Pradesh	1.25	3.98
Nagaland	0.17	4.08
Manipur	2.21	3.34
Mizoram	0.99	5.67
Tripura	2.43	3.51
Meghalaya	1.71	5.13
Assam	4.09	3.67
West Bengal	6.29	2.83
Jharkhand	2.05	1.67
Orissa	3.92	3.32
Chhattisgarh	1.69	2.46
Madhya Pradesh	2.49	2.60
Gujarat	1.82	3.26
Daman & Diu	0.05	4.99
D & N Haveli	0.03	1.62
Maharashtra	4.73	3.40
Andhra Pradesh	7.23	2.04
Karnataka	4.12	2.36
Goa	0.52	3.80
Lakshadweep	0.41	5.43
Kerala	10.77	5.63
Tamil Nadu	7.59	3.17
Pondicherry	0.28	6.16
A&N Islands	0.64	3.14
India	**54.7**	**2.95**

Source: Author

Table 11.5 provides state-wise distribution of female-headed households (percentage) and their average education level (mean schooling years) in rural India. On average, 54.7% of female-headed households have an average 2.95 or nearly 3 years of education. The state having the highest number of female-headed households is Kerala (10.77%) with 5.63 years of education, while Delhi has the least female-headed households with maximum 10.48 years of education.

The weighted mean education level of female-headed households in different states of rural India is also computed (Table 11.5). On the basis of the national average education level, we observe that some states are above the national average education level, while some states are below it. Now, we compare the educational level of female-headed households with high firewood and chips consuming states. We observe that generally, those states where the average education level of the female-headed households is below the national average level are high consumers

Table 11.6 States above and below national average education of female-headed households in rural area

States above mean educational level of female-headed households		States below mean educational level of female-headed households	
States	Weighted mean educational level	States	Weighted mean educational level
Jammu & Kashmir	3.56	Haryana	2.37
Himachal Pradesh	4.86	*Rajasthan*	1.96
Punjab	4.04	*Uttar Pradesh*	2.30
Chandigarh	4.32	*Bihar*	2.41
Uttaranchal	3.37	*West Bengal*	2.83
Delhi	10.48	*Jharkhand*	1.67
Sikkim	4.20	Chhattisgarh	2.46
Arunachal Pradesh	3.98	*Madhya Pradesh*	2.60
Nagaland	4.08	D & N Haveli	1.62
Manipur	3.34	*Andhra Pradesh*	2.04
Mizoram	5.67	*Karnataka*	2.36
Tripura	3.51		
Meghalaya	5.13		
Assam	3.67		
Orissa	3.32		
Gujarat	3.26		
Daman & Diu	4.99		
Maharashtra	3.4		
Goa	3.80		
Lakshadweep	5.43		
Kerala	5.64		
Tamil Nadu	3.17		
Pondicherry	6.16		
A & N Islands	3.14		

Source: Author's calculation

of firewood and chips. Jharkhand, Rajasthan, Bihar, Uttar Pradesh, Andhra Pradesh, etc. belong to this category and are reported in green colour in Table 11.6.

We have also considered the land-holding pattern of households in the determination of fuel choice focusing on land owned, land possessed and land cultivated. Since our focus is on rural areas and choice of firewood and chips, we compare the state-wise mean land holdings of owned, possessed and cultivated with high firewood and chips–consuming states. It is observed that there are some land-rich states which are high consumers of firewood and chips like West Bengal, Bihar, Jharkhand, Tripura, Uttar Pradesh and Rajasthan. These states are poor states according to our estimate except Rajasthan. There are some land-poor states who are high consumers of firewood and chips such as Andhra Pradesh, Maharashtra, Karnataka, Tamil Nadu and Gujarat, but they all belong to the category of non-poor in our estimates.

From preliminary data analysis, we observed that only 4091 (64.13%) female-headed households participated in firewood and chips selection out of nine alternative available cooking fuels in rural India (Table 11.1). As per response, in the first stage, 4091 rural female-headed households gave the response 'yes' for firewood and chips selection and the rest (2288) gave the response 'no'. However, in the second stage, 3637 rural female-headed households consumed a positive amount of firewood and chips for their cooking purposes. So, 454 rural female-headed households participated in firewood and chips selection in the first stage, but they actually consumed 'zero' amount of firewood and chips in the second stage. Hence, there is a difference between desired and actual consumption of firewood and chips due to certain factors. Here, we apply the double hurdle model for determining factors as it is discussed in methodology parts in Chap. 3.

11.3 Determinants of Firewood and Chips Consumption of Female-Headed Households in Rural India

In this context, Cragg (1971) put forward a two-part extension to Tobin's model and addressed the presence of two latent variables through two regression equations, i.e. participation decision and actual consumption (or expenditure) decision. The estimated participation and quantity equations are reported in Table 11.7, which displays the result of the double hurdle model (DHM) for the selection of firewood and chips and its actual consumption decision of female-headed households in rural India.

Table 11.7 presents the estimated results of participation and quantity equations for the choice of firewood and chips and consumption decision considering households' assets like land ownership (LO). The results of the participation equation (Table 11.7) show that demographic features like family size and age of household head and its square; education (average educational level of household), secondary and above education level of household head; social factors like castes ST, SC, and OBC; and Hindu and others group are statistically significant. These variables

Table 11.7 Estimated results of firewood fuel choice and consumption (DHM) in female-headed households in rural India

	Participation equation		Quantity equation	
Variable name	Coefficient	t-value	Coefficient	t-value
Constant	1.359	0.76	0.656	0.51
LogMPCE	−0.254	−0.55	**0.941****	**2.80**
LogMPCE_sq	0.0042	0.14	**−0.051****	**−2.31**
LO	−5.65	1.59	**−11.4*****	**−4.78**
Age	0.0051	0.67	0.0066	1.30
Age_sq	−0.000026	−0.30	**−0.0002*****	**−3.67**
Hhsize	**0.0616*****	**4.31**	**0.064*****	**7.6**
Edu	**−0.072*****	**−5.61**	**0.0175****	**1.99**
Age_head	**0.043*****	**4.97**	**−0.019****	**−2.84**
Age_head_sq	**−0.0004*****	**−4.98**	**0.00023*****	**3.8**
Edu_head				
Below primary	0.0577	0.84	**0.105****	**2.3**
Primary	0.01197	0.17	**0.1232****	**2.57**
Middle	0.0828	1.09	0.0635	1.21
Secondary	**−0.2486****	**−2.92**	0.087	1.33
Higher secondary	**−0.36*****	**−3.16**	0.0415	0.45
Graduate	**−0.392****	**−2.81**	0.1454	1.21
Postgraduate and above	**−0.82*****	**−3.67**	−0.28	−1.24
Diploma/certificate course	−0.357	−1.49	**−0.444****	**−2.08**
Through TLC	0.29	0.45	**−0.648****	**−2.01**
Social group				
ST	**0.34*****	**5.08**	0.08	1.83
SC	**0.306*****	**5.13**	**−0.182*****	**−4.48**
OBC	**0.28*****	**6.24**	**−0.241*****	**−7.47**
Religion				
Islam	0.017	0.29	0.0687	1.74
Christianity	0.04	0.51	0.0653	1.22
Others	**−0.415*****	**−4.86**	−0.0137	−0.20
N=	6375			
Sigma	0.927			
Covariance	**−0.7616*****	−28.56		

Source: Author
Note: '***', '**' and '*' denote 1%, 5% and 10% level of significance, respectively

determine the choice of firewood and chips as cooking fuel of female-headed households in rural India. However, the amount of consumption of firewood and chips is determined by quantity equation in the second stage. Statistically significant factors are income and its square, Land owned (LO), family size, average educational level of household, age of household head and its square, and household head's educational levels at below primary, primary, diploma, technical certificate course, etc. (see Table 11.7).

The estimated results of the participation equation suggest that income is not a determining factor for selecting firewood and chips consumption in rural India; however, the quantity equation suggests that the amount of firewood and chips consumption rises with income at a declining rate. These findings indicate that income has no significant role in the selection of firewood and chips in female-headed households in rural India; whereas income crucially determines its actual quantity consumption, if it is selected.

Positive coefficients of household size in participation and quantity equations indicate that an increase in household size increases the probability of choosing firewood and chips, and its amount of consumption also rises. It suggests that the rising manpower of households probably assists in collecting more firewood and chips in rural areas.

Land ownership (LO) may provide insightful probable information regarding the choice of fuel consumption, particularly firewood and chips, at the household level in rural areas. LO is statistically insignificant in the participation equation. So, land has no role in firewood and chips selection. The negative coefficient of LO is highly significant in the quantity equation. This suggests that the amount of firewood and chips consumption declines with land holding. This finding indicates a possibility to switch over to alternative fuels accessible and/or available in rural India.

The positive coefficient of "age of household head" and the negative coefficient of its square in the participation equation suggests the probability of choice of firewood and chips increases with a diminishing rate along the age of household head. The signs of coefficients of age of household head and its square are just reverse in quantitative equation referring to participation equation. So, firewood and chips quantity consumption declines with age of female-headed households, but it rises at older age.

Negative coefficient of education in participation equation represents a declining probability of firewood and chips selection with rising average education of household, while its positive coefficient in quantity equation indicates that if households consume firewood and chips, they make positive amount of consumption. These results might be due to the lack of access to cleaner or alternative fuels; truly, it is difficult to comment anything on it without further study.

The educational level of female-headed households plays a crucial role in the selection of firewood and chips and its consumption amount. Negative coefficients of secondary, higher secondary, graduate and postgraduate education of female-headed households in the participation equation indicate that the probability of selection of firewood and chips decreases with secondary and higher levels of education of female-headed households. The quantity equation reflects a slightly different picture. The positive coefficient of primary and below primary is significant and indicates the amount of firewood and chips consumption rises in primary and below primary educated female-headed households. However, the negative coefficients of diploma and technically educated female-headed households are quite surprising which indicate the reduction in the consumption of firewood. It is noted that a desirable reduction of firewood and chips consumption is required for achieving at least a secondary level of education of female-headed households in rural India. Hence, policy should be designed to raise the educational level of the girl child beyond the secondary level and ensure the availability of clean fuels matching with demand associated higher level of education.

The double hurdle model presents the effect of different social groups, namely ST, SC and OBC, compared to others on the fuel selection behaviour of households. Positive coefficients of ST, SC and OBC groups in the participation equation indicate that the probability of choice of firewood and chips rises for female-headed households of ST, SC and OBC compared to others. So, the probability of dirty fuel selection increases in female-headed households of socially deprived castes. The positive coefficient of ST and negative coefficients of SC and OBC in the quantity equation reflect that compared to others, ST households consume more firewood and chips, while SC and OBC households consume it in lesser amount. For religions, in the participation equation, it is observed that only other groups have a chance to reduce the probability of selection of firewood and chips.

11.4 Conclusion

This chapter raises the research issue of hurdles of fuel choice of female-headed households in rural India. This study identifies the determining factors of fuel choice and its non-zero quantity demand at rural female-headed household levels focusing on hurdles in fuel choice decision at different stages in India. The results reflect significant demographic and socio-economic variables like family size, education, social castes and educational level of female-headed households. Income and land are not determining factors of dirty fuel like firewood and chips selection in rural India. The choice of firewood and chips in rural India basically depends on family composition, education of female heads and social castes, whereas consumption incidence depends on the said factors along with income and land.

This study highlights the importance of land ownership as an important factor for determining the asset structure of rural female-headed households in rural India. Firewood and chips, being a dirty fuel, has several ill effects and such fuel consumption should be discouraged. Results indicate that an increase in the education of household heads at least up to the secondary level lowers the probability of selection of firewood and chips and reduces its consumption. Hence, policymakers should focus on policy design for improving female education beyond the secondary level. Empirical findings suggest to shift some consumers of firewood and chips to other clean fuels through the educational attainment of female or raising awareness regarding dirty fuel consumption. Female participation should be encouraged at decision-making and its execution that may help shifting households from the consumption of dirty to clean fuels. The Government of India has initiated it through Ujjwala gas scheme for rural poor. It is true that women suffer directly from ill effects of dirty fuel consumption and so their voice in decision-making of households might bring about some positive change in fuel consumption patterns of rural households in India.

Chapter 12
Conclusions and Policy Recommendations

12.1 Introduction

This study has raised an important research issue the determinants of domestic fuel choice decision at household level. Fuel choice of households and their actual consumption decision under their socio-economic conditions is required to understand for achieving Sustainable Development Goals (SDG) within 2030. Households' fuel choice and amount of fuel consumption decision might certainly help to control the risk of uncertainty associated with climate change which is a potential threat to sustainable development in emerging economies like India. This study has attempted to identify the determining factors of cooking fuel choice decision at the household level under various socio-economic conditions that definitely help to reformulate and redesign energy policy focusing on targeted SDGs. To make an effective energy policy design for achieving SDGs, we have to identify factors influencing households' fuel choice decisions in two main stages: the first stage is the selection of cooking fuel from alternatives, and the second stage is the quantity consumption decision of selected fuel at the first stage.

12.2 Summary of the Major Findings

Fuel choice decision crucially depends on non-price factors in rural areas of developing economies, like India. Rural households in India mostly prefer firewood and chips to the rest of the nine fuels. The major contributing chapters have discussed the determinants of fuel choice and consumption using limited dependent models. However, the desired fuel consumption is different from their actual cooking fuel consumption being zero due to certain non-price factors. There might be some households who report a positive desire of consumption of cooking fuel, but their

S. Daripa, S. Dinda, *Hurdles in Fuel Choice in Rural India*, SpringerBriefs in Geography,
https://doi.org/10.1007/978-3-031-71605-8_12

actual consumption of that cooking fuel is zero. There might be certain factors that inhibit carrying out of the change for actual consumption to be positive. The tobit model assumes that participation and consumption decisions can be modelled as a single equation, whereas Cragg (1971) relaxed this assumption and models both of these as separate regression equations introducing the double hurdle model. The double hurdle model is applied to overcome such limitation in this study that determines the choice of firewood and chips consumption in rural India. This model reduces observational loss compared to probit and truncated regression models.

This chapter also makes a comparative analysis of the double hurdle model by excluding and including land assets in the model. A comparative study indicates that incorporating the effect of land-rich households in this study gives a clear picture about the determinants of firewood and chips consumption in rural India and hence highlights the importance of land ownership as an important factor for determining the asset structure of rural households.

Firewood and chips, being a dirty fuel, has several bad effects, and such fuel consumption should be discouraged. Our results indicate that an increase in the education of household heads at least up to the secondary level reduces the probability of choosing firewood and chips, as well as its consumption amount. Hence, policymakers should focus on policy design for improving education for all, especially female education beyond the secondary level. Empirical findings suggest shifting some firewood and chips consumers to other clean fuels through the educational attainment of household heads or raising awareness regarding dirty fuel consumption. Female participation should be encouraged at the household-level decision-making that may help shifting households from consumption of dirty to clean fuels. It is true that women suffer directly from the ill effects of dirty fuel consumption, and so their voice in decision-making of households might bring about some positive change in the fuel consumption patterns of rural households in India.

This study observes the variations of determinants for bottom and top quartiles, largest populated state (UP), castes and also sex of the household heads. Ultimately, the double hurdle model provides better results than the probit and truncated regression models.

This study has discussed the choice of cooking fuel across castes in rural India. The study has identified various demographic characteristics such as family size, average age of the family and age and sex of the family head; economic factors like income and land; and social factors like level of education of family as well as head of the family, religious beliefs, etc., which are highly associated with consumption decision for firewood and chips across castes (scheduled tribes (ST), scheduled caste (SC) and other backward classes (OBC)) in rural India. The majority of rural female-headed households of deprived castes (ST, SC and OBC) are mostly landless and illiterates, and they have nearly no option except firewood and chips freely available in rural areas. Households having land in all castes reduce their consumption of firewood and chips. Land ownership rights might ensure reduction of dirty fuel such as firewood and chips consumption.

12.3 Policy Recommendations

The fuel source might play an important role in taking several policy decisions regarding fuel choice at the household level in a developing country like India. Empirical findings suggest shifting some firewood and chips consumers to other clean fuels through educational attainment of household heads or raising awareness regarding dirty fuel consumption. Female participation should be encouraged at the household-level decision-making that may help shifting households from consumption of dirty to clean fuels. It is true that women suffer directly from the ill effects of dirty fuel consumption, and so their voice in decision-making of households might bring about some positive change in the fuel consumption patterns of rural households in India.

Advancement of the economic position of the households changes their attitude towards dirty fuel like firewood and chips selection as well as consumption. Discrete policy choice will be more effective than uniform fuel policy. Recently, the government of India has adopted a policy like the Ujjwala gas scheme for rural poor and around more than 8 crore households are benefited from the said programme. It should be noted that the policymakers should be given more focus on accessibility and availability of alternative clean fuels for all section of rural households especially old age and female-headed households in rural areas.

The policymakers may also think about educational attainment and rural land distribution. Asset reallocation policy should be assigned with carbon accounting and green credit that definitely ensures Sustainable Development Goals (SDGs). So, the government should redesign land distribution policy for the deprived social castes across rural India and ensure targets of SDGs within 2030.

Policymakers should focus on policy design for improving female education beyond the secondary level and ensure the availability of clean fuels matching with demand-associated higher level of educations.

12.4 Concluding Remarks

The desire and actual fuel consumption of the households are generated in two different processes associated with socioeconomic conditions. Cragg (1971) model integrates two different behavioural experiences in his double hurdle model, which provides the outcome of two different processes corroborating socioeconomic situations. This study has certain limitations. The intercept term in the participation equation is still highly significant which indicates omission of some variables that should be incorporated in the future study. Education, in this study, is considered as a categorical variable; however, the results could have been different if we could have used it as a continuous year of schooling. We have also not considered the interaction effects of some variables like education and sex of the household head which might help in further analysis of firewood and chips consumption pattern in rural India. The effect of such interaction terms could give us important insight for effective policy design and further research in this arena.

References

Adrianzen, M. A. (2013). Improved cooking stoves and firewood consumption: Quasi-experimental evidence from the Northern Peruvian Andes. *Ecological Economics, 89*, 135–143.
Akenji, L., & Bengtsson, M. (2014). Making sustainable consumption and production the core of the sustainable development goals. *Sustainability, 6*(2), 513–529.
Andreae, M. O. (1991). Biomass burning: Its history, use and distribution and its impact on environmental quality and global climate. In J. S. Levine (Eds.), *Global biomass burning: Atmospheric, climatic and biospheric implications* (pp. 3–21). MIT Press.
Arnold, J. E. M., Kohlin, G., & Persson, R. (2006). Woodfuels, livelihoods and policy interventions: Changing perspectives. *World Development, 34*(3), 596–611.
Banerjee, M., Prasad, R., Rehman, I. H., & Gill, B. (2016). Induction stoves as an option for clean cooking in Rural India. *Energy Policy, 88*, 159–167.
Barnes, D., Khandker, S., & Samad, H. A. (2010). *Energy access, efficiency and poverty: How many households are energy poor in Bangladesh?* (Policy Research Working Paper. WPS5332). The World Bank.
Brouwer, R., & Falcao, M. P. (2004). Wood fuel consumption in Maputo, Mozambique. *Biomass Bioenergy, 27*(3), 233–245.
Campbell, B. M., Vermeulen, S. J., Mangono, J. J., & Mabugu, R. (2003). The energy transition in action: Urban domestic fuel choices in a changing Zimbabwe. *Energy Policy, 31*(6), 553–562.
Couture, S., Garcia, S., & Reynaud, A. (2012). Household energy choices and fuelwood consumption: An econometric approach using French data. *Energy Economics, 34*, 1972–1981.
Cragg, J. G. (1971). Some statistical models for limited dependent variables with application to the demand for durable goods. *Econometrica, 39*, 829–844.
Daripa, S., & Dinda, S. (2020). Sustainable consumption pattern in India. In S. Hazra & A. Bhukta (Eds.), *Sustainable development goals* (Sustainable Development Goals Series). Springer. https://doi.org/10.1007/978-3-030-42488-6_12
Daripa, S., & Dinda, S. (2022). Hurdles in fuel choice and consumption in rural India. In S. Bagli, G. Chakrabarti, & P. Guha (Eds.), *Persistent and emerging challenges to development* (India Studies in Business and Economics). Springer. https://doi.org/10.1007/978-981-16-4181-7_6
Daripa, S., & Dinda, S. (2024). Stage-based non-price determinants of fuel choice and its consumption decision of households in rural India. *The Indian Economic Journal*, (forthcoming). https://doi.org/10.1177/00194662241238589
Davis, M. (1998). Rural household energy consumption: The effects of access to electricity – Evidence from South Africa. *Energy Policy, 26*(3), 207–217.
DeFries, R., & Pandey, D. (2010). Urbanization, the energy ladder and forest transitions in India's emerging economy. *Land use policy, 27*(2), 130–138.

S. Daripa, S. Dinda, *Hurdles in Fuel Choice in Rural India*, SpringerBriefs in Geography,
https://doi.org/10.1007/978-3-031-71605-8

Gasper, D., Shah, A., & Tankha, S. (2019). The framing of sustainable consumption and production in SDG 12. *Global Policy, 10*(1), 83–95.

Gebreegziabher, Z., Mekonnen, A., Kassie, M., & Kohlin, G. (2012). Urban energy transition and technology adoption: The case of Tigrai, Northern Ethiopia. *Energy Economics, 34*, 410–418.

Goswami, A., Bandyopadhyay, K. R., & Kumar, A. (2017). Exploring the nature of rural energy transition in India: Insights from case studies of eight villages in Bihar. *International Journal of Energy Sector Management, 11*, 463.

Gundimeda, H., & Kohlin, G. (2008). Fuel demand elasticities for energy and environmental policies: Indian sample survey evidence. *Ecological Economics, 30*, 517–546.

Gupta, G., & Kohlin, G. (2006). Preference for domestic fuel: Analysis with socio-economic factors and rankings in Kolkata, India. *Ecological Economics, 57*, 107–121.

Hanna, R., & Oliva, P. (2015). Moving up the energy ladder: The effect of an increase in economic well-being on the fuel consumption choices of the poor in India. *The American Economic Review, 105*, 242–246.

Heltberg, R. (2004). Fuel Switching: Evidence from eight developing countries. *Energy Economics, 26*, 869–887. https://doi.org/10.1016/j.eneco.2004.04.018

Heltberg, R. (2005). Factors determining household fuel choice in Guatemala. *Environment and Development Economics, 10*, 337–361.

Hosier, R. H., & Dowd, J. (1987). Household fuel choice in Zimbabwe: An empirical test of the energy ladder hypothesis. *Resources and Energy, 9*, 347–361.

Jingchao, Z., & Kotani, K. (2012). The determinants of household energy demand in rural Beijing: Can environmentally friendly technologies be effective? *Energy Economics, 34*, 381–388.

Jumbe, C. B. L., & Angelsen, A. (2011). Modelling choice of fuelwood source among rural households in Malawi: A multinomial probit analysis. *Energy Economics, 33*, 732–738.

Kebede, E., Kagochi, J., & Jolly, C. M. (2010). Energy consumption and economic development in Sub-Sahara Africa. *Energy Economics, 32*, 532–537.

Leach, G. (1992). The energy transition. *Energy Policy, 20*(2), 116–123.

Ma, Y., Rong, Ke., Luo, Y., Wang, Y., Mangalagiu, D., & Thornton, T. F. (2019). Value co-creation for sustainable consumption and production in the sharing economy in China. *Journal of Clean Production, 208*, 1148–1158.

Masera, O. R., Saatkamp, B. D., & Kammen, D. M. (2000). From linear fuel switching multiple cooking strategies: A critique and alternative to the energy ladder model. *World Development, 28*(12), 2083–2103.

Narasimha Rao, M., & Reddy, B. S. (2007). Variations in energy use by Indian households: An analysis of micro level data. *Energy, 32*, 143–153. https://doi.org/10.1016/j.energy.2006.03.012

Nayak, B. P., Werthmann, C., & Aggarwal, V. (2015). Trust and cooperation among Urban poor for transition to cleaner and modern cooking fuel. *Environmental Innovation and Societal Transitions., 14*, 116–127.

Ouedraogo, B. (2006). Household energy preferences for cooking in urban Ouagadougou, Burkin Faso. *Energy Policy, 34*(18), 3787–3795.

Pachauri, S., & Jiang, L. (2008). The household energy transition in India and China. *Energy Policy, 36*(11), 4022–4035.

Raman, P., Ram, N. K., & Murali, J. (2014). Improved test method for evaluation of bio-mass cookstoves. *Energy, 71*, 1–17.

Scherer, L., Behrens, P., de Koning, A., Heijungs, R., Sprecher, B., & Tukker, A. (2018). Trade-offs between social and environmental sustainable development goals. *Environmental Science and Policy, 90*(1), 65–72.

Smith, K. R., & Mehta, S. (2003). The burden of disease from indoor air pollution in developing countries: Comparison of estimates. *International Journal of Hygiene and Environmental Health, 206*(4–5), 279–289.

Swarup, V. A., & Rao, K. R. (2015). An econometric approach to analysis of trends and patterns of household fuel choices in India. *Indian Economic Review, 50*(1), 105–129.

Takama, T., Tsephel, S., & Johnson, F. X. (2012). Evaluating the relative strength of product-specific factors in fuel switching and stove choice decisions in Ethiopia. A discrete choice model of household preference for clean cooking alternatives. *Energy Economics, 34*, 1763–1773.

Tobin, J. (1958). Estimation of relationships for limited dependent variables. *Econometrica, 26*(1), 24–36.

van der Kroon, B., Brouwer, R., & van Beukering, P. J. H. (2014). The impact of the household decision environment on fuel choice behaviour. *Energy Economics, 4*(1), 236–247.

Wooldridge, J. M. (2008). Chapter 17: Limited dependent variable models and sample selection corrections. In *Introductory econometrics: A modern approach* (4th ed., pp. 574–622). South-Western College Publication.

Xu, S., Chu, C., Zhang, Y., Ye, D., Wang, Y., & Ju, M. (2018). Entangled stakeholder roles and perceptions of sustainable consumption: An evaluation of sustainable consumption practices in Tianjin, China. *Journal of Environmental Management, 223*, 841–848.

Index

S. Daripa, S. Dinda, *Hurdles in Fuel Choice in Rural India*, SpringerBriefs in Geography,
https://doi.org/10.1007/978-3-031-71605-8

The manufacturer's authorised representative in the EU is Springer Nature Customer Service Centre GmbH, Europaplatz 3, 69115 Heidelberg, Germany. If you have any concerns regarding our products, please contact ProductSafety@springernature.com

Printed and bound by CPI Group (UK) Ltd, Croydon, CR0 4YY

30/09/2026

02225474-0004